HOW TO **MAXIMISE** YOUR WEALTH OFFSHORE

This book is dedicated to

Emily and Katherine

HOW TO **MAXIMISE** YOUR WEALTH OFFSHORE

DEBORAH BENN

B. T. BATSFORD LTD · LONDON
Batsford Business Online: www.batsford.com

© B T Batsford 1998
First published 1998

Published by:
B T Batsford Ltd,
583 Fulham Road,
London SW6 5BY

Printed by
Redwood Books
Trowbridge
Wiltshire

ISBN 0 7134 8380 6

A CIP catalogue record for this book
is available from The British Library.

CONTENTS

INTRODUCTION   9

CHAPTER ONE
SELECTING AN OFFSHORE CENTRE   11
Why go offshore?   11
What is an offshore centre?   13
Is it legal?   15
Is it for you?   17
Case studies   19
Factors when selecting an offshore centre   22
Tips when selecting an offshore centre   24
Where to start   25

CHAPTER TWO
COMMUNICATION AND THE INTERNET   27
Review of offshore financial websites   28

CHAPTER THREE
TAX   37
Tax planning: Dispelling the myths   37
Tax status of Investors and Products   38
Taxation of offshore products   40
Domicile and Residence status   41
Tax Treaties   41

Treaty Shopping   42

Countermeasures   44

Avoidance versus Evasion   45

CHAPTER FOUR

FINANCIAL ADVICE   49

Tips on sound financial advice   49

What to look for in a financial adviser and the costs involved   50

What if things go wrong?   52

CHAPTER FIVE

TRUSTS   55

What is an Offshore Trust?   55

The Difference between Common law and Civil law   59

Trust Definitions   59

Different types of Trusts   63

Reasons for establishing a Trust   65

Factors when Selecting an Offshore Trust Jurisdiction   67

CHAPTER SIX

COMPANY FORMATION   71

What is an Offshore Company?   71

Reasons for establishing an Offshore Company   72

Regulation and costs   73

Different types of companies and their uses   74

Tips on establishing an Offshore Company   78

Factors when selecting an Offshore Company jurisdiction   79

CHAPTER SEVEN

BANKING   83

Deposit Accounts   87

Multi-currency Accounts   88

Private Banking   89

Mortgages   91
Share dealing   92

CHAPTER EIGHT
INVESTMENT MANAGEMENT USING OFFSHORE PRODUCTS   97
Types of Products Available   98
Product Profiles   100

CHAPTER NINE
OFFSHORE FUNDS   105
Explanation of the different offshore fund structures   107
Fund pricing   111
Fund solutions   113
Roll-up Funds   115
Hedge Funds   119
Offshore fund performance   125
The impact of charges on fund performance   133
How to pick a winning offshore fund   138

CHAPTER TEN
OFFSHORE CENTRE INFORMATION   149
Introduction   149
Bahamas   151
Barbados   153
Bermuda   155
British Virgin Islands   157
Cayman Islands   162
Cook Islands   166
Cyprus   169
Dublin   174
Gibraltar   176
Guernsey   179
Isle of Man   183

Jersey   187

Liechtenstein   191

Luxembourg   194

Netherlands Antilles   198

Switzerland   200

Turks & Caicos   204

Further centre contacts   206

GLOSSARY   209

USEFUL ADDRESSES   221

INDEX   223

INTRODUCTION

As access to offshore opportunities increase via new technology such as the Internet, so will the need for specific information on the advantages - and the pitfalls - of investing offshore.

How to Maximise your Wealth Offshore strives to address this increasing use of offshore centres by a wider network of investors. Whether you have investments offshore already or plan to in the future, it is designed to give you the tools you need to make necessary choices enabling you to invest offshore with confidence. It not only dispels some of the myths surrounding this investment medium, but also highlights the often-complex arrangements that can be part and parcel of investing offshore and which should not be stepped into without a clear understanding of the consequences.

The emphasis has been placed on de-jargonising the offshore terms you are likely to meet. Case studies have also been used to highlight the different offshore options available to investors in a variety of circumstances. The Internet site review of a selection of offshore investment groups gives a good flavour of the fast moving pace of this communication mode.

Rather than re-invent the wheel, *How to Maximise your Wealth Offshore* utilises the knowledge of experts and includes a number of articles from well-known players in the offshore field. I would like to thank the following for their contributions:

Paul Forsyth of Forsyth Partners
Richard Timberlake of Portfolio Fund Managers
David Wells of Arthur Andersen
Michael Lipper and Christina Romas at Lipper Analytical, US
Giles Corbin and Steven Meiklejohn at Ogier & Le Masurier, Jersey
Victor Hills, adviser to Argyll Investment Management
Guinness Flight Asia
Albany International, Isle of Man

In addition, I would like to thank on-line information provider, Bloomberg for much of the basic research for the book, particularly the information on Offshore Centres. Also, Moneyfacts and Old Mutual International for contact and tabular information and, finally, Kira Nickerson for her enthusiastic website review.

Deborah Benn
May 1998

CHAPTER 1

SELECTING AN OFFSHORE CENTRE

WHY GO OFFSHORE?

Investing offshore is often associated with millionaire status and powerful institutions. But increasingly, there are opportunities open to a wider variety of investors. According to fund performance analysts, Lipper Analytical, retail investors now hold 90% of total offshore fund assets compared to only 4% held by institutions. The reasons for the large number of retail investors can be attributed to a number of factors. One factor is global mobility of the workforce. Take the example of a British expatriate couple working in the Middle East or a Dutch scientist working in a war zone - different sets of nationalities, different types of employment - the common factor being that they are often outside their normal country of residence and need a central home for their finances. A further factor is the vast number of domestic investors who invest in offshore funds to legally avoid withholding tax - this is a common feature in Europe, where, say, German banks will set up an operation in low tax Luxembourg simply to target investors back in Germany.

It is this desire to protect and preserve assets which is the catalyst to explore the investment opportunities offshore. By investing offshore you can protect your assets from high taxes or economic instability which can severely erode their value. In countries where political instability generally puts assets at risk, investing in a 'neutral' or safe haven is one way of protecting your assets.

Most offshore activities could be described as tax planning, using legal mechanisms to reduce or eliminate taxes on income, wealth, profits and inheritance. While minimising current and future taxes remains an attractive incentive, the growth in popularity of offshore centres is also due to the lack of investment restrictions that exist there. This lack of restrictions allows companies to offer more innovative investment products and services. This is particularly true of offshore funds whose numbers have grown rapidly due in part to the investment freedom they have compared with onshore funds. There are now more than 6,000 offshore-based funds to choose from. The chapter on offshore funds explains about the choice of funds on offer and takes a look at the best performing funds identified by Lipper Analytical.

Furthermore, it is the ability to offer solutions for an ever-changing investment audience that has seen offshore centre use becoming more mainstream. Offshore centres are now at the cutting edge of new corporate, investment, trust, insurance, partnership and banking legislation and are among the first jurisdictions to offer unique structures such as limited partnerships, asset protection trusts, purpose trusts and limited duration companies.

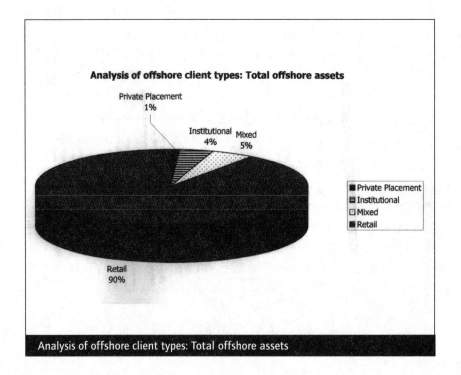

Analysis of offshore client types: Total offshore assets

WHAT IS AN OFFSHORE CENTRE

Before looking at the factors behind selecting on offshore centre, it is important to clarify what exactly an offshore centre is. In its simplest form, an offshore financial centre is a jurisdiction that offers a low or no tax regime, as well as a high level of confidentiality. Sometimes called tax havens, these jurisdictions exist around the world and offer a 'neutral' or safe haven for investments.

Additional benefits include no exchange control restrictions and a wide variety of tax efficient vehicles such as trusts, companies and investment products. It was the search for a neutral home for investments that became the catalyst for the modern offshore centre. At the time of the Second World War, social and political turmoil in and around Germany, Russia, and South America forced many people to move their assets to the stability of neutral Switzerland. Not only was Switzerland neutral, it also provided the anonymity of secret, numbered bank accounts. Following the end of the war, the need to raise taxes to fund massive reconstruction programmes encouraged many people to continue using the relatively low taxed Switzerland.

Switzerland's success in offering a low tax and confidential home for assets seeking political and social refuge led to a number of jurisdictions declaring themselves as 'offshore centres'. The number of offshore centres in existence has increased dramatically. Comprehensive details of a selection of these offshore centres are listed in the final chapter. However, highlighted points from each centre are listed below.

In order to gain a competitive edge, offshore centres are specialising in banking, trust and company formation or offshore funds. For example, offshore financial centres include six of the world's 12 top banking centres involved in corporate finance and individual investments. These include Cayman Islands, Luxembourg, Hong Kong, Switzerland, the Bahamas and Singapore. Luxembourg deals mostly in offshore funds and is now the fourth largest investment funds centre in the world. The Cayman Islands is another big offshore fund centre. The total value in offshore funds is now more than US$1 trillion and is beginning to rival the US domestic mutual fund industry in size.

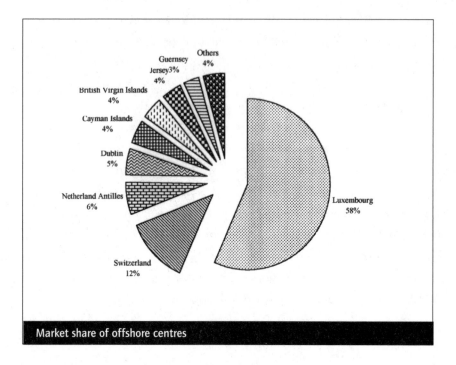

Market share of offshore centres

Along with the tax breaks, freedom from excessive investment restriction and regulation is the reason why banking, insurance and investment have become important pillars of the offshore industry. Offshore centres frequently permit diversification into investment activities not allowed at home. This means they can often use more innovative investment structures such as derivatives or loan facilities or access more exotic investment markets. While this can increase the risk, it is also designed to enhance returns. Such a 'flexible investment system' is not to be confused with a slack regulatory regime, which allows companies of dubious background whose prime purpose is to target unsophisticated investors with poor offshore investment plans. While this no doubt still goes on, reputable offshore centres have very comprehensive regulatory regimes and the newer centres are continually improving their regulations.

Indeed, many of the world's top banks and financial institutions have established operations in offshore centres. For example, more than 40 of the world's top 50 banks have operations in the Cayman Islands alone. Household names which have an offshore presence include Barclays Bank, Halifax, Fidelity, Perpetual, Rothschild, Chase Manhattan, Citibank, Goldman Sachs, Charles Schwab, Clerical Medical and Scottish Widows to name but a few. In addition, many of these groups will have a presence in more than one centre. Many centres have adopted the policy of allowing only blue chip companies to establish operations, to discourage the less scrupulous establishing operations to target private investors. Examples of such centres include Jersey and Guernsey. As well as checking the policy of centres in allowing groups to establish there, it is worthwhile checking out the groups that have already set up an operation in order to gauge whether or not the offshore centre regards itself as a serious player.

IS IT LEGAL?

The offshore financial industry is often portrayed as a home for illegal enterprise and tax evasion. This notion is misleading and outmoded. While fraud no doubt goes on, offshore centres do not have a monopoly on it, inasmuch as it happens equally onshore.

Through reciprocal relationships and compliance with domestic investment regulations, certain offshore funds are marketed freely to the public in most European, British Commonwealth and, by special agreement, in most industrialised countries in Asia, including Japan. Public offshore funds are unable to be sold in the US because their sponsors choose not to go through the cumbersome Securities and Exchange Commission (SEC) registration process.

In fact, it is perfectly legal for most people to invest offshore. What is illegal is not to declare to the tax authorities any income or gains you receive from your investments. If you currently reside outside your normal country of residency - you may be working abroad - then the rules to some extent depend on your nationality and residency status. For example, subject to qualifying tax rules, a

British expatriate has the ability to use an offshore centre without having to declare any income or gains to the UK Inland Revenue so long as he or she has been out of the UK for 5 years. Obviously, you may still have to declare any income or gains to the tax authorities where you currently reside, but as an expatriate you may be subject to more flexible rules than indigenous residents.

Not all nationalities are subject to similar privileges. US citizens, for example, have to declare their worldwide income and gains no matter how long they are resident outside the US. Although this does not preclude them from investing in offshore funds, the fact that each individual share held within a fund has to have income and gains accounted for separately makes it a complicated process.

The tax rules and regulations governing the use of offshore centres and residency can in most cases be obtained from your tax authority.

It is worthwhile bearing in mind that domestic rules governing offshore investments are subject to change. The March 1998 Budget in the UK outlined certain changes applicable to offshore investments by UK residents and non-residents. The impetus for such changes is generally because of a 'loophole' which exists and gives investors an unnecessary advantage in the eyes of the law. For example, one of the changes in the UK's March 1998 Budget was to close the 'dead settlor' loophole, which allowed offshore trusts to escape taxation completely if the settlor died.

In terms of fraudulent use of offshore centres, authorities worldwide are cracking down heavily. Practically all offshore jurisdictions have concluded Mutual Legal Assistance Treaties with the US, UK, Canada and other countries to formalise the prevention of illegal transactions in order to stamp out money laundering. New laws are being passed to criminalise the misuse of offshore financial institutions for such illegal purposes.

The definition of a 'criminal' transaction presents a major stumbling block for offshore centres. Specifically, whether or not tax evasion is considered a criminal offence. While not wanting purposely to impede justice, this notion attacks the essence of every offshore centre in that tax evasion cannot be considered a crime in a country where there are no taxes. It is therefore hard to justify financial disclosure if a crime committed abroad is not considered a crime in the haven country. Most centres have a Bank Secrecy Code, which prohibits all bank personnel from divulging information about a client's account to any outside source - such as a court, and/or tax collector. The penalties are severe for divulging such information. Most banks do not consider it their responsibility to act as another government's revenue service.

However, in an effort to show authorities that centres are not being used to launder money and conscious that tax authorities can make life difficult for them, most offshore centres have now made certain concessions in the spirit of stamping out "dirty" money. The degree of concessions varies. In the case of Switzerland this has gone further than expected by allowing tax authorities access to information where they can prove that tax evasion has taken place.

IS IT FOR YOU?

The offshore financial services industry is in transition from institutions serving the very wealthy and large international businesses to a much broader-based clientele of middle market corporate and private customers attracted to a wide choice of broad-based fiduciary, trust, brokerage, and other investment services. Offshore financial institutions are now actively seeking the business of a much broader range of investors, and the formation of offshore trusts and offshore companies is soaring.

The decision to utilise an offshore financial centre is one that obviously must be given careful consideration. If tax avoidance is the primary motivation, then somewhere along the line you are likely to come under the scrutiny of tax authorities that are continually tightening up on such abuse. If, on the other hand, your goal is to access new markets, take advantage of business and investment opportunities unavailable at home or to legally protect assets from damaging legislation, then the use of an offshore financial centre offers a suitable number of options.

However, coming to a decision is not as easy as it sounds. Despite offshore-based institutions seeking business from a much broader base, it is not commonplace to hear about opportunities to invest offshore. It is therefore sometimes difficult to make an informed decision one way or the other. There are sometimes valid reasons for this lack of information on investing offshore. These include:

–   **Investor Protection:** As a general rule, it is illegal to promote an offshore product to the onshore public unless the product has been authorised by the domestic regulators. For example, in the US offshore funds have to be authorised by the SEC in order to be able to be promoted within the US. In the UK, offshore funds have to be authorised by the Financial Services Authority (FSA). Once a fund is FSA recognised it can be marketed freely within the UK.

–   **Domestic tax incentives:** Onshore investment products sometimes attract tax relief, which minimise the attractiveness of offshore products. Wherever possible, you should utilise any onshore investment tax relief before opting to go offshore.

–   **Not suitable for the mass market:** Offshore products may be aimed at specific sectors of the market, such as high net worth individuals or institutions. This makes mass marketing of information irrelevant.

In most cases it is up to you to find out about the opportunities. Due to the sometimes complicated nature of investing offshore, it is important to always use the services of a qualified professional adviser who has offshore expertise. The section on seeking financial advice deals with this in more detail. Also below are some specific examples of situations where offshore investment is suitable or unsuitable.

However, there is no substitution for being aware of the pros and cons yourself. At the very least, it will ensure you are fully armed with the correct information when talking to your adviser. It is also worth while keeping track of domestic developments that could present an offshore window of opportunity. For example, the ending of a tax benefit for investors in Swedish-domiciled investment funds at the beginning of 1997 saw just such an opportunity for offshore funds. Prior to this, investors were liable to capital gains tax at 20 per cent, as opposed to the 30 per cent levied on the returns from offshore funds. Now the capital gains tax is a uniform 30 per cent for both onshore and offshore funds, which has levelled the competitive playing field.

Tax changes are not the only developments to keep an eye on. Distribution is another area to monitor. Taking the example of Sweden again, since 1995 insurance brokers have been permitted to act as financial advisers, allowing them to distribute mutual funds which hitherto could only be sold through banks.

CASE STUDIES

One of the best ways to demonstrate whether or not investing offshore is for you is to take a look at some case studies written by the experts. The following case studies were compiled by **David Wells, financial planning manager at Arthur Andersen**. In all four case studies, the non-UK tax position of the investors has not been considered. Obviously, this would need to be taken into account. Arthur Andersen has a dedicated department on international tax and investment advice.

Case 1

British expatriate couple, working in the Middle East with 2 school age children. Wife does not work. He is well paid and in five years they plan to return home to the UK, where they have property.

As non-UK residents, this couple will only be taxed on income that arises in the UK. Also, they will escape the UK capital gains tax regime. As a result, it may prove advantageous, subject to the prevailing tax rules in the Middle East country of residence, to invest offshore. By doing so, all liability to UK tax will be avoided so long as money is not remitted to the UK.

As UK nationals, the couple should primarily invest in equity funds with a UK bias, as ultimately sterling is their base currency. If they were overseas on 17 March 1998, they should, from a tax perspective, consider selling any UK assets heavy with capital gains shortly before return to the UK, in order to realise the gains while non-resident. Those leaving the UK to become non-resident after 17 March 1998 will, if they return within five complete tax years, be taxable in the UK in the year in which UK residence is resumed on any gains realised in their absence. As the wife does not work, UK assets could be held in her name to take advantage of her personal allowance and lower tax bands. If she does invest offshore, however, the money will benefit from gross roll-up of income, which will be paid without the deduction of UK tax credits. She could buy UK gilts.

Case 2

US married couple both working in the UK on three-year renewable contracts. Plan to start a family but do not plan to return to the US for some time. Have the possibility of working in another European country.

This couple will be classed as resident from the date of arrival in the UK and as such will be liable to income tax on both UK and, on a remittance basis only, overseas income. In addition, they will be liable to tax on any gains made whilst resident in the UK.

Investing in UK and offshore collective funds is complicated for US nationals by the US tax filing requirement which stipulates that each individual share held within the trust has to have income and gains accounted for separately. Consequently, it is simpler for US nationals to invest in recognised US mutual funds or directly into equities.

The mutual funds in which they invest should primarily be geared towards their home currency as a hedge against any adverse movements in the US dollar against the pound.

Case 3:

British retired couple. Empty nest. Plan to spend retirement in Spain and have a reasonable amount of savings. Want to keep the property in the UK as they plan to return frequently to see children and grandchildren.

As non-UK residents, this couple will only be taxable in the UK on income that arises in the UK. With regard to capital gains tax, the new rules announced in the March 1998 Budget should not affect them if they are resident in Spain for five complete tax years or more.

Given that they will be living in Spain, they should look to invest in offshore funds - thus escaping UK income tax and capital gains tax - with a bias towards the peseta and other key European currencies, including the Euro. Depending on their income requirements, their portfolio should have a reasonably high equity content to maintain its real value and provide a growing income. They might also hold eurobonds and UK gilts.

This couple should note that unless they sever all ties with the UK; they will retain their UK domicile status and hence be liable to UK inheritance tax on their worldwide assets. They must be careful that their visits to the UK are not sufficient to make them resident there.

Case 4:

Two high fliers in the media. He is British, she is American. Heading for divorce. Husband is in line to receive a mini fortune on conversion of options in the business he helped set up, which is about to be listed on the stockmarket. No children.

As UK residents, they will be liable to income tax on both UK and overseas income. As the wife is not domiciled in the UK, however, her liability to UK income tax on overseas income will only arise on a remittance basis. In addition, they will be liable to tax on any gains made while resident in the UK.

It will probably be virtually impossible to invest overseas in an attempt to shelter money from the divorce settlement. The courts will undoubtedly look through any such transactions and factor the money back into the equation. The exact divorce settlement will be subject to the length of their marriage and in which jurisdiction (UK or US) the divorce is being filed. The couple should note that the UK courts now have the capability of 'earmarking' any pension provision which either spouse has accumulated, if it is deemed that the other's provisions are relatively inadequate.

The husband could consider taking advantage of the reinvestment relief available for capital gains, which effectively defers the capital gains tax bill. To make this work, he will have to reinvest the gains from his options in new shares of an unquoted company that conducts a 'qualifying' trade or a venture capital trust. The main exclusions from the definition of a qualifying trade are financial services and property backed companies. This transaction must take place within a period of one year before to three years after the exercise of the options (one year after in the case of reinvestment in a venture capital trust).

FACTORS WHEN SELECTING AN OFFSHORE CENTRE

The choice of a suitable offshore centre is basic to modern international tax planning. The selection of an offshore centre will depend on a number of factors. Not least, if you have already identified a specific product and provider then the centre in which the provider is based will determine the choice of centre. Major factors when it comes to choosing an offshore centre include:

Service: There are great variations between services rendered in different offshore jurisdictions, and it is no longer an easy matter to decide the best venue to serve a particular financial need. The degree of specialisation will play a part. For example, Luxembourg and Dublin are at the forefront of offshore investment funds, the British Virgin Islands in the formation of International Business Companies, Jersey in offshore trusts, the Caymans and Bahamas in offshore banking.

Geographic location and focus: The Caribbean centres cater particularly to US and Canadian needs. Those in the Channel Islands and British Isles relate more to British Commonwealth countries. Cyprus accommodates East European needs, Mauritius focuses on Indian Ocean trade, and Dublin and Madeira attempt to act as gateways to the European Union. Dealing with people in the same time zone is also a benefit, although telephone banking and Internet use is breaking down these barriers. (See Offshore Centre information).

Tax efficiency: Not all offshore financial centres levy zero taxes. A country that levies tax based on source only will be a tax haven in respect of income derived from foreign sources. Similarly, a country whose tax base is restricted to its residents will be a haven in respect of income accruing to non-residents.

Other considerations:

- political and economic stability

- confidentiality

- specific laws governing companies, trusts and other entities

- the speed and ease of establishing, running and winding up a trust or company

- guarantees against future taxes (the Cayman Islands government offers attractive guarantees to both exempted trusts and exempted companies)

- zero tax treatment of non-resident and foreign income

- flexibility - certain centres provide that the domicile of a company may be moved out of the jurisdiction in the event of a change of circumstances

- the existence of a tax treaty network or double taxation agreements (certain centres owe their existence to these networks)

- any exchange controls

- the existence of anti-tax avoidance legislation/government attitude to tax haven operations

- Banking and professional facilities must be adequate for the needs and intentions of the operation.

TIPS WHEN SELECTING AN OFFSHORE CENTRE

- Reputation - don't be fobbed off with platitudes. Ask around yourself and speak to people who have used an offshore centre.

- Are regulations tight and standards of professionalism high? Asking which companies have been refused a license can be a more telling gauge than who is already based there.

- Does the centre specialise in a particular type of investment or service? If it is not the type you are after, then you should look at alternative offshore centres which do specialise in that particular area and will likely be more geared-up to your specific needs.

- How insulated is the centre from the laws of foreign jurisdictions, such as forced heirship rules? Has banking secrecy been pierced and under what particular circumstances?

- Does the centre's regulatory body operate under a code of conduct? Is it a member of any international group that can uphold its reputation?

- Is there legislation to protect investors?

WHERE TO START

- Get informed. Try and get hold of specific offshore financial publications (see useful addresses). Unfortunately, most specialist publications can only put you on the mailing list once you have already moved abroad - which defeats the object of pre-planning. This is down to domestic regulations and not the fault of the publisher. If you are currently an expatriate or planning in the near term to become an expatriate either working or living abroad then you should be able to get yourself on the mailing lists.

- Contact domestic investment groups you currently favour and find out whether they have any offshore operations, asking them to send details of what they can offer. Many do excellent explanatory leaflets and guides explaining the principles of offshore investment.

- Contact the offshore centre regulatory authorities (see offshore centre information). They often publish booklets or guides for investors.

- Contact your local tax authority to obtain specific information or guidelines on the tax treatment of offshore investments.

- If you do plan to work abroad through a company then the Human Resources department will be able to supply you with information and make recommendations.

CHAPTER 2

COMMUNICATION AND THE INTERNET

It is perhaps the remote distance involved in dealing with an offshore centre that makes it a daunting prospect for many investors. Many offshore-based companies operate in different time zones from the investor and some parts of the world start and end their weekend at different times from the offshore location. Public holidays and religious festivals can all contribute to the frustration and difficulty of being able to get access to your investments when it is needed most.

For this reason excellent communication is vital. Technology is playing a vital role in helping to eliminate some of these difficulties and enabling information to be made available instantly in a more cost-efficient way.

The good news is that most offshore banks, trust companies and fund management and life groups are gearing up to the communications challenge by creating 24 hour telephone centres and establishing sites on the world wide web. A selection of these offshore financial sites is reviewed on the following pages. Not only does a website provide a show case for the company's products and services, it can also give you instant information on a company *before* you deal with it.

Some companies are more advanced than others in the electronic facilities they offer. While some can offer the facility to answer requests for information, others give you the ability to form a company over the Internet.

This is certainly an area set to increase, but one that should be taken with great care. As a relatively new form of communication it is still unregulated.

According to the Centre for the Study of Financial Innovation (CSFI), an independent London-based think tank committed to exploring the future of the finance industry, it believes the Internet poses new challenges to the cyber police. In its report, The Internet and Financial Services, it highlights areas such as fraud, software piracy, system contamination, money laundering and hacking and the problems of inconsistency of regulation around the world in dealing with such Internet crime. In Europe, the Distance Selling Directives (there are currently two versions) will incorporate regulation on financial services and the Internet. The directives are currently in draft form.

While methods of regulation and safe electronic payment are currently being developed, the safest option for the time being is to stick to big names when using this method of investing.

REVIEW OF OFFSHORE FINANCIAL WEBSITES

In today's world of expanding technology, investors are spoiled for choice in the way in which they can receive information and invest their money. These days the majority of offshore fund groups now allow instant access to detailed information on their company and funds, and some even allow investors to trade via the Internet without leaving their computer. Kira Nickerson, of City Financial Communications, runs through a few of the more popular websites.

GT Global, located at **www.gtglobal.com**, has a flowery home page outlining the company's awards and top funds. The site offers basic company information. From the home page, the browser can enter the market reviews section that offers a library of articles on market reviews. The global overview, was nearly a month out of date when the site was reviewed (on 16 March 1998), however, the weekly bond report was more recent. The site offers prices and historic performance on GT Global funds using Standard and Poor's Micropal information. You can view prices for week to date; last week; fund performance tables 1-10 years; and 1-6 months. The remaining section gives a menu in which readers can select literature they would like sent to them concerning specific funds. Underlying this is the company address and contact numbers. Overall the graphics were overpowering. However, due to its simple format and the basic information available, the site is easy to access and surf through.

Hansard Group, the offshore life assurance company, can be accessed at **www.hansard.com** Clicking on the Hansard International button will take international investors to the portion of the site containing the most pertinent information for their interests. The options listed are limited, but informative. Background is available on the company, which informs first time browsers where the company is located, gives information on tax exemptions available from the offshore centre site located in the Isle of Man and provides general company history.

The product key gives a quick overview of Hansard's products, what the minimum contributions are, currencies and more. The unit prices link takes you to a table that shows the fund, the percentage change that week and the current prices. The final option is a contact list, which gives the company address, telephone and e-mail.

While the site is simple and easy to manoeuvre for even the most computer-shy investors, there's very little more here that you would not be able to read in a company brochure. The graphics are colourful and the overall look of the site is busy but overdone, considering the amount of information contained.

Without a doubt the most informative site, with the most up-to-date technological tricks, is **Charles Schwab**. The US site, at **www.schwab.com** is good for background information of what Charles Schwab can offer and features a demo programme showing how net users can make stock transactions via the web. The information contained here is so extensive that those unfamiliar with the net could be easily intimidated. For those who want simply to examine what it can offer, the best decision would be to use the site map key located at the bottom of the site home page.

While the US site is interesting, the Schwab web site that would affect non-US resident offshore investors is located at **www.schwab-worldwide.com**. This site, equally as informative but less intimidating, features details on Schwab investments in Asia and the Cayman Islands. Check out the mutual fund centre in the Asian section, where browsers can view funds by family, category or look up a specific fund. In addition to performance, investors can read a detailed profile, which includes an overview of the funds' distribution method, inception, total assets, minimum investment and the fund manager. The profile also looks at the fees and expenses and the average annualised total return.

The site has even more information that can only be accessed by account holders. Those with a Schwab account, by entering their code and account number, can enter a secure area and read company reports, Dow Jones news, plus access information on their accounts and investments, look at their portfolios, and make transactions.

The company even provides support through help-lines, toll-free customer service and direct contact links on the site. The worldwide site is simple to use and understand and moves quickly between screens.

Global Asset Management, at **www.gam.com** is another attractive web site packed with information on the company. The sharp looking home page offers browsers a quick look at the company, a fast link to the company's top six funds for 1998, today's prices, frequently-asked questions and a world map with GAM locations. The sight of a face behind the company, Chairman Gilbert de Botton, adds a personal touch as well as an easy link to his biography so viewers can learn more.

From there the next choice could be the top six funds which, following the disclaimer on investment risks, moves to the top six GAM funds that the company believes will perform well this year as well as their current performance and details.

The 'today's prices' were disappointing when viewed. Surprising for such a forward-looking web site, the prices listed were two days out of date. (Viewed 6 March with 'daily' prices listed for 4 March.)

The features section allows viewers to receive email updates on the company and adds a link to the *Financial Times*. Also, it offers the most frequently-asked questions, such as why cannot a US citizen invest in offshore funds?

The site also appears to allow registered users to trade online. However, all the site offers is an application form for the desired fund which, when completed, must be mailed to the GAM.

Old Mutual, located at website **www.oldmutual.com** has a busy looking home page brightly decorated with a map providing direct links to the company's sites throughout the world. Tucked in amidst the colourful photos is a brief company description and a menu that includes a search engine, contact, company biography and site guide.

The site guide is almost a repeat of the home page, offering the same services. However, overall it is confusing and contains little or no real help. In order for browsers to find Old Mutual International they need to go to the site guide and follow the tiny printed menu down until it appears mid-way through a complicated list of options which seems similar to the fine print of a contract. Clicking on Old Mutual International transports you to information on international investment services. The information under Which Jurisdiction is fairly sparse but useful. Other useful elements include a brief description of the different types of offshore investment vehicles offered by Old Mutual International, a checklist of investment goals plus fund information.

By clicking on one of the countries located on the home page map, viewers are transported to that office, or so it would seem. Instead, visitors are given a brief description of the office located in that country. It resembles international yellow pages for Old Mutual, which most people could obtain by telephoning the London office or the international operator. A short cut to clicking each individual place would be to click on one and then hit the button called 'our companies' on the right-hand menu. This then lists a brief description of all the company's branches.

Fund and unit prices are also an option from each individual country site that accesses the same information and is updated weekly. Recent addition to the sites has been a new South African site. However, while one has been added, another has been taken off line during the time the North American site undergoes construction.

Opening with the very business-looking design of a filo-fax style home page, is **Baring Asset Management's** site, at **www.baring-asset.com**. Offshore investors can read what offshore services the company offers by clicking on the filo-fax page tab to the left of the screen. Once past the Internet red tape of disclaimers, you arrive at a page listing the services but not the funds. For fund information, visitors must backtrack through red tape once again after clicking the mutual funds button. At this site there is a selection of options including corporate structure, offices, investment philosophy, range of funds, a forum and daily prices. By selecting the Dublin-based offshore funds section of the menu, apparently viewers are able to see the daily prices which the site states: "is updated on a daily basis, and published here on the site." However, nothing appeared when I tried the site on 16 March 1998.

The rest of the information on the entire site is simply where to find the company, how to contact it and its investment philosophy. While the site is quick and easy to negotiate, with the exception of the numerous disclaimers, it holds little more than a company brochure.

An example of a well designed, easy-to-use, useful and informative international site, both in terms of company and market information, is **Banque Internationale à Luxembourg** at **www.bil.lu**. The home page appears in English but is also offered in German, Dutch and French. The options menu allows visitors to look at a range of things from SICAVs to exchange rates to news and general market information. When users click on the SICAV key, they are taken to a sub menu offering financial markets and reports in addition to risk management. The design of this page allows a good understanding of the issue without even reading the information below. The colourful pyramid outlines the different areas where investors can put their money and places them in ascending order of risk.

Another key feature to this sight is Bilonline - a web trading service. Those seeking what it can offer can run the demo and see that by using this programme investors are able to examine their portfolio in real time and check out their accounts and security deposits. They are also able to carry out, through the easy-to-use electronic form, stock transactions and transfers.

The rest of the information is like most sites with clear, concise investment news examining the company's services from its advantages to its consulting and shareholders' services.

While investors cannot use the **Guinness Flight** site, located at **www.guinness-flight.com** to invest their money directly, it does offer more than other offshore fund web sites. Visitors or interested investors click how to invest from the main menu, then select a fund listed in the fact sheet index page. After reading the funds' key features with details including dealing times and charges, they can print off an application form and mail it, with payment, to the company.

Choices are abundant in the funds index with the option to look at anything from money funds to multi-currency funds. Each of these again leads to daily prices, asset allocation information and a monthly portfolio review.

Other information the site contains is a daily fund prices page, funds index, and performance statistics. The features section includes bonds, a guide to roll-up funds, investing in Asia and other general market information.

Under the fund key features section is a list of frequently asked questions. While it looks as if each individual question has a link to the answer, in fact if any question is clicked the viewer will be shown the entire list of questions and answers.

Despite the fact that the site has many useful options listed, there are some drawbacks. The colour choice on the home page makes reading about the company virtually impossible. In fact, the only way of knowing something is written there is by using your mouse to highlight the seemingly empty screen. In addition, there are many links leading the visitor around the site but because so many of them overlap, it becomes more difficult to negotiate than other sites.

Below is a selection of offshore sites where investors can access company, fund and stockmarket information plus offshore centre sites.

www.flemings.lu
www.fidelity.com
www.nation-ltd.co.uk

Name of Centre	Web site
Bahamas	www.interknowledge.com/bahamas/investment
Barbados	www.barbados.org/business.htm
Bermuda	www.bsx.com/cgi_win/bermuda
British Virgin Islands	www.bvi.org
Cayman Islands	www.cimoney.com.cy
Cook Islands	www.ck/offshore.htm
Cyprus	www.kypros.org/cbc/offshore
Dublin	www.ida.ie/fs.htm
Gibraltar	http://www.gibraltar.gi/fsc
Guernsey	http:gfsc.guernsey.net
Isle of Man	www.isle-of-man.com/
Jersey	http://www.itl.net/business/ofw/Jersey
Liechtenstein	n/a
Luxembourg	www.alfi.lu/
Netherlands Antilles	www.gov.an/min2.html
Switzerland	www.giradin.org/switzerland/
Turks & Caicos	www.offshoreweb.com/bbp_index/financial_servi ces_commission

Table 3

Did you know?

- More than 600 banks around the world already have web sites

- Digital signatures are rapidly acquiring legal status

- An Internet bank's running costs are one third those of traditional banks

- 1.2 million US investors have on-line stockbroking accounts

- Many UK web sites are very probably non-compliant with SIB and Serious Fraud Office (SFO) regulations

Source: Centre for the Study of Financial Innovation

CHAPTER 3

TAX

TAX PLANNING: DISPELLING THE MYTHS

It is probably best to start this chapter off by dispelling a few myths. First, when you invest offshore, it does not necessarily mean you pay less tax or no tax at all. Take the example of someone investing in an offshore fund located in an offshore centre where there are no income or gains taxes. What this means is that the fund will not be taxed, which enables it to grow tax free and pay dividends gross. This compares with most onshore investment rules that tax the fund at source. This tax-free growth should then feed through to increased offshore fund performance. However, as an investor in the fund, it is quite possible you may personally suffer a tax liability depending on where you are resident and your current tax status.

So using an offshore centre does not absolve you from paying tax. What the use of an offshore centre does give you, apart from tax-free growth of your investments, is tax planning opportunities that should lead to paying less tax. Generally speaking, you will not incur a tax liability until you either cash in your investment or repatriate it onshore, and this is where the planning comes into the equation.

The extent of freedom you have for tax planning and the amount of tax you save depends on a number of factors. These factors include whether you are an individual resident in your normal country of residency, an expatriate, a business owner or someone planning to live abroad sometime in the future. Each country has its own tax rules and regulations relating to offshore investments. For example, US citizens have little offshore tax planning opportunities as they are taxed on their worldwide income irrespective of whether it is repatriated or cashed in. Anomalies arise due to the fact that some countries such as South Africa and Hong Kong assess tax on a source basis only, while most Western countries assess tax on a residence or citizenship basis, or a combination of all three. However, it is not within the scope of this book to outline each individual country's tax rules relating to investing offshore. It is important, therefore, that you seek guidance from your own tax authorities and consult with a qualified tax planner or financial adviser with tax knowledge. It addition, before you take advantage of any offshore tax breaks, you should fully utilise any onshore tax breaks to the full.

TAX STATUS OF INVESTORS AND PRODUCTS

1. **An individual resident in their normal country of residence.** This probably gives the least flexibility for tax planning and depends to a large extent on the size of the assets you have to invest. However, there are opportunities worth exploring. Taking the UK as an example, one tax planning benefit is the ability to defer paying tax until you are, say, in a lower tax band or near to retirement.

2. **An individual planning to retire abroad.** Often the fact of emigration provides an individual with the ideal opportunity for international tax planning using offshore centres. This is mainly down to the fact that moving from one country to another gives you an element of freedom in your choice of fiscal system. This allows ample opportunities to invest in one or more offshore centres prior to settling in your chosen destination. It is worthwhile noting the laws on domicile and residence. When moving from one country to another, the fact of domicile becomes important in international tax planning. See section on domicile and residence below.

3.   **An expatriate.** This status gives you a good degree of flexibility in
      terms of tax planning and saving tax. In some countries, expatriates are
      accorded a different tax status than the residents. Expatriates resident
      in the UK, for example, are taxed on the 'remittance' basis. That is they
      are taxed only on income and gains derived within the UK. The amount
      of flexibility you have as an expatriate will depend on the length of
      time you are away from your normal country of residence and how
      much time you spend in other countries while working. For example, a
      UK expatriate has to stay abroad for five years before they are not liable
      to UK income and capital gains tax on their investments. Return to live
      in the UK within five years and you will be taxed in the year you return
      on your investment income. This change was made in the March 1998
      UK Budget.

4.   **Internationally mobile employees.** Executives working for companies with
      overseas bases, entertainers and sportsmen often have to work abroad.
      Double taxation agreements and other advantageous foreign tax laws
      mean tax savings can be achieved by structuring the remuneration package
      of the executive so as to minimise taxation on that remuneration.
      However, tax authorities are becoming aware of abuses in this area. The
      March 1998 UK Budget stopped the foreign earnings deduction (FED)
      whereby substantial tax relief on any income was given if it was
      demonstrated that the income arose in a foreign country for services
      rendered in that country.

5.   **A business owner.** Once a business starts to expand internationally, often
      the lack of compatibility between different countries' tax regimes can give
      rise to anomalous situations, which make an international business
      operation too expensive to develop. An extreme example could then see
      the business paying overseas taxes as well as taxes in the home country.
      There is often an advantage in routing dividends, interest or royalties into
      a centre which imposes some taxes, but which enjoys the benefit of a
      double taxation agreement that reduces the rate of tax at source. This

may result in a lower aggregate tax charge than using a zero-tax centre, which in certain countries will attract a high rate of tax. See Treaty Shopping on page 42.

TAXATION OF OFFSHORE PRODUCTS

Some offshore products have been especially developed to take advantage of certain tax planning strategies available in a number of countries. The most common ones are:

Offshore deposit accounts: These accounts pay interest gross, giving taxpayers the advantage of not having to pay tax until after they have withdrawn the interest. Compare this with the UK where 20 per cent tax is deducted at source. (Although UK non-taxpayers can get interest paid gross). However, if you are UK resident and invest in an offshore deposit account then you must still declare this to the Revenue, even if you do not withdraw the interest. You will eventually have to pay tax. The main advantage to UK resident taxpayers in offshore deposit accounts is an increase in cashflow.

Roll-up Funds: There are two main types of offshore funds. 'Accumulator' funds that roll up income and gains on investments within the fund each year, and 'Distributor' funds which pay out income and gains. While distributor funds can put you in the position of having to pay tax in your country of residence, with roll-up funds there is generally no tax to pay until you cash in all or part of your holdings. This is a useful tax deferral benefit, particularly if you plan to move or work abroad or anticipate falling into a lower tax bracket. It is worth noting that although you are subject to tax on distributor funds, you can use any onshore tax relief to offset this. You cannot use onshore tax relief for roll-up funds.

Insurance bonds: These have similar advantages to roll up funds in that gains are accumulated within the bond. Additionally, if you are UK resident you can withdraw five per cent of your investment each year without having to pay any tax until you cash in the bond. It is also possible to switch between investment sectors within the bond without triggering a tax charge.

DOMICILE AND RESIDENCE STATUS

Put simply, an individual's domicile is the country that he or she regards as their permanent home. Thus, if you become resident in a country, but still tend to return to your previous home at some stage in the future, your domicile will not coincide with your residence. Movement from one country to another, even for a relatively long 'temporary' absence, is unlikely to change an individual's domicile. This factor may be of prime importance because certain fiscal systems levy tax on the basis of domicile, in that persons who are resident but not domiciled in the particular country receive certain tax concessions. By the same token, someone who has lived in another country for many years but dies without changing domicile could see the domicile tax authorities also having a claim on the estate. It is very difficult to change domicile. Despite having lived abroad for many years, if someone keeps even the smallest connection - a forgotten bank account, for instance - then it can always be proved the person had the intention of returning.

TAX TREATIES

The Basics

Tax treaties can be complicated to follow but are useful when it comes to international tax strategies. The following information supplied by on-line information provider, Bloomberg, highlights some of the tax treaty basics as well as the principles of treaty shopping.

Countries form double-taxation agreements (treaties) to prevent or reduce international double taxation on the same income (of an individual or corporation). Understanding the special benefits available in these agreements and between which nations they exist are important factors that should be considered in international investment, trade, or services.

Generally, income tax treaties prevent double taxation by allowing just one of the countries to tax a particular item(s) of income, such as dividends, interest, royalties, earned income, based on certain basic principles. In some cases, however, both countries are given the right to tax. The right of taxation may be given to the country:

1.   that is the source of income,

2.   of which the recipient is a resident; or

3.   of which the recipient is a citizen. When both countries have the right to tax under these principles, double taxation may be eased through a foreign tax credit allowed by the country of citizenship or residence, for example the US.

To achieve these purposes, tax treaties contain a wide variety of provisions including tax exemptions, reductions in tax rates, tax deductions or credits, rules on allocation of income and expenses between related taxpayers, and provisions for the exchange of information.

TREATY SHOPPING

International investors and corporations will often attempt to exploit the most favourable tax treaty arrangement by 'treaty shopping'. Treaty shopping is the act by third country residents to benefit from the advantages of a treaty.

The usual procedure is for the 'shopper' to establish or relocate a corporate entity in a treaty partner's jurisdiction to benefit the owner located elsewhere. Switzerland, with its extensive treaty networks, Cyprus, with its treaty ties to the developing nations of Eastern Europe, and Barbados, with its favourable treaty with the US and its large market, are all frequently used in treaty shopping.

Name of Centre	Exchange Controls	Tax Treaties	Taxes
Bahamas	Nil	Nil	No withholding, capital gains, inheritance taxes.
Barbados	No (For IBCs)	Several	Offshore financial sector subject to low taxes. No capital gains or inheritance tax.
Bermuda	Nil	Nil	No withholding, capital gains, transfer, wealth, gift or income taxes for non-resident entities
British Virgin Islands	Nil	2	Zero taxes for offshore entities
Cayman Islands	Nil	Nil	Zero taxes
Cook Islands	Nil	Nil	Zero taxes
Cyprus	Nil	Many	Offshore companies pay tax.
Dublin	Nil	Yes	Non-resident companies are taxed solely on income from Ireland
Gibraltar	Nil	Nil	Non-resident owned and controlled companies incorporated in Gibraltar which do not trade, earn or remit income to Gibraltar are not liable to company tax.
Guernsey	Nil	UK, Jersey	Exempt companies are only taxed on local source income.
Isle of Man	Nil	UK	Non-residents are only taxed on local suorce income
Jersey	Nil	UK, Guernsey	No capital gains, estate or inheritance taxes.
Liechtenstein	Nil	Austria	There is no tax levied against any company that is domiciled
Luxembourg	Nil	Many	No withholding tax on bank or bond interest and certain holding companies are exempt from tax on profits
Netherlands Antilles	Nil	Few	Offshore income from dividends and interest is taxable. Income from capital gains is tax exempt
Switzerland	Nil	Many	Tax rates vary greatly among the cantons. No withholding tax on interest income from foreign bonds issued in Switzerland
Turks & Caicos	Nil	Nil	Zero taxes

TAX Tables Sheet 1

Mainly people and companies with extensive international interest use it. Put simply: by establishing a company structure in various offshore centres that have treaty networks, you can legally exploit the double tax treaty agreements to ensure you are not taxed twice. By careful centre selection you can choose the lowest rates of tax. Obviously, this is a complicated area and tax authorities are mindful that it is open to exploitation. For this reason, the steps taken need to be genuine in order for the structure to be watertight.

It is also worthwhile noting that in an effort to stamp out flight capital, higher taxed industrial countries have denied withholding tax reductions if the true beneficiary was not a resident in one of the treaty nations. This has made treaty shopping more difficult. Treaties are also being renegotiated to include clauses that stipulate an exchange of information between nations.

COUNTERMEASURES

The success of offshore centres has not gone unnoticed. By last count there were over 25 tax havens worldwide and there always seems to be a "new kid on the block". The relatively high-tax industrial nations are keen to stem this increasing outflow of potential revenue by developing their own stratagems for counter-attack. Even with existing treaties, high tax countries have denied withholding tax reductions if the true beneficiary was not a resident in one of the treaty nations. Treaties are also renegotiated to include clauses that stipulate an exchange of information between nations. Because the US taxes its individual and corporate residents on all income from all worldwide sources, it has found it necessary to develop the most extensive information exchange network with other countries. The UK is not far behind.

Treaty or no treaty, there are indeed limits to the extent one nation will surrender information or grant access to a foreign government. As mentioned earlier, some jurisdictions will only compromise confidentiality if it concerns criminal activity indictable in both countries. Moreover, an exchange of information is not always perceived to be equitable and beneficial to both parties. Notice that very few no-tax jurisdictions have tax treaties with high tax nations. Some other measures employed by industrialised nations such as France and the US to discourage tax avoidance and evasion include legislation limiting offshore tax deferral schemes, imposing exorbitant fines and penalties, and rewarding informers. In the year ended September 30, 1996, the US IRS paid 650 rewards totalling about US$3.5 million, down from the previous year of 671 rewards, but up from US$1.8 million paid. The IRS claims tipsters helped it collect $102.7 million in taxes, fines, penalties and interest in 1996, up from $96.4 million in 1995. The record year was 1993, when the IRS paid $5.3 million to informants.

AVOIDANCE VERSUS EVASION

Tax avoidance is the legal escape from taxes through provisions allowed in the tax code. Evasion, however, is the illegal non-payment of taxes either by overstating deductions and or exemptions, or by under-declaring income. However, the lines between evasion and avoidance are often blurred even for the professionals. By far the most reliable method to ensure you are not crossing the law is to look at legal precedent. Legal precedents are tax cases that have already gone through the courts and are referred to by the courts when assessing fresh cases. The following article by **Giles Corbin and Steven Meiklejohn** at Jersey-based legal firm Ogier & Le Masurier sets out the current thinking.

As governments (particularly the UK) redefine what constitutes tax avoidance and tax evasion, how can investors be sure of their actions?

The judiciary has in recent times complicated and confused the previously clear distinction between legal tax avoidance and illegal tax evasion. Investors must stay in touch with these developments in order to avoid finding themselves on the wrong side of the law. This article attempts to summarise the traditional approach to the distinction between avoidance and evasion.

Traditionally there have been two fundamental concepts. Tax avoidance, involving the reduction of the incidence of tax borne by an individual taxpayer contrary to the intentions of Parliament where the reduction did not involve concealment. Tax avoidance in this sense was lawful. Tax evasion, on the other hand, involved a concealment of the facts and has obviously been criminal. Typically, acts deceiving the Revenue or failing to inform authorities of a liability where there is a duty to do so constitute evasion.

However, there has been a blurring of the clear distinction between legal tax avoidance and illegal tax evasion by virtue of a sub-division of the definition of tax avoidance into legal and illegal strains.

Lord Templeman has innovated the phrase "tax mitigation" to describe the tax avoidance acts that constituted legal tax avoidance. This has resulted in a trinity of concepts: tax evasion, tax mitigation (acceptable tax avoidance) and tax avoidance (of an unacceptable character).

Income tax mitigation is where a taxpayer reduces the income or incurs expenditure in circumstances that reduce his assessable income or entitle him to a reduction in his tax liability. For instance, where a taxpayer makes a settlement, he deprives himself of the capital that is a source of income and thereby reduces his income. If the settlement is irrevocable and satisfies certain other conditions the reduction in income reduces the assessable income to the taxpayer. The tax advantage results from the reduction of income that is an act of legal tax mitigation. This confirms the general observation that there is no duty upon individuals liable to resident taxation to maximise their local tax liability.

This new terminology has probably resulted from the discomfort that the judiciary and particularly Lord Templeman felt with the then wider definition of tax avoidance. In Fitzwilliam versus IRC he stated ..."I regard tax avoidance schemes of the kind invented and implemented in the present case as no better than attempts to cheat the Revenue."

It is recommended that anybody using offshore tax planning structures studies the case of R versus Charlton (1996) STC 1418 very carefully. The approach of the Inland Revenue, the conduct of the case and the comments of the Court of Appeal should be of concern to all.

There are two types of arrangements in question in this case. The first involved UK persons ordering goods from overseas through offshore companies specially set up for them. The second involved new offshore companies being paid for consultancy services rendered to UK persons. There is nothing wrong with such arrangements so long as, in the first case, the offshore companies genuinely order the goods and they are then sold on to UK persons; and, in the second case, the offshore companies provide genuine services for the money they receive. Whether either scheme will actually save any tax is quite another matter. There is some particularly tough anti-avoidance legislation, the purpose of which is to nullify the tax advantages of such arrangements. Nevertheless, a properly set up and implemented offshore arrangement, even if the tax savings are nullified by the anti-avoidance legislation, is neither fraud nor evasion.

The introduction of concepts of 'acceptability' and 'fairness' within R versus Charlton are unhelpful on the basis that if a person's conduct is illegal, then it is right that he is punished. But if his conduct is legal, the only question should be: 'Does the tax law impose a tax liability. If it does he should - save for situations where extra statutory concessions apply - pay the tax and, if the law does not impose a tax liability, then he should escape tax.

However, the jury in this criminal case appears to have been confronted by the wrong question. In our view the revenue incorrectly focused on the question: 'were the transactions bona fide commercial ones?' when in this matter it seems to us guilt or innocence could not possibly turn on this. We believe the jury should have been asked: 'were the transactions real (regardless of whether or not they were bona fide commercial ones)?'

Another revenue submission was based on the facts that the directors of the UK company continued to negotiate the purchases, fixing the price of each consignment and making the necessary arrangements for direct delivery. Most significantly, the United Kingdom directors instructed the Jersey company what price it was to charge for the goods in each transaction. The only explanation the Revenue could see for such an arrangement was the creaming-off of part of the profits made by the UK company so that less corporation tax could be paid and funds would be built up offshore for the benefit of the UK directors. In answer to this, it should have been said that the "creaming-off" of part of the profits to reduce corporation tax is not a criminal offence. On the contrary, the only way to cream-off profits is to enter genuine transactions, not bogus ones.

R versus Charlton, which has been described as a blot on our system of jurisdiction by an eminent tax QC, has some important lessons for those who seek to use low tax jurisdictions and legitimate tax planning. It is imperative that such arrangements must be properly carried out, and that no corners must be cut. Not only must things be done properly, but also they must unequivocally be seen to be done properly.

CHAPTER 4

FINANCIAL ADVICE

TIPS TO SOUND FINANCIAL ADVICE

Successful utilisation of tax treaties, tax codes, various corporate entities, investment vehicles and financing techniques is considered by some to be an art form. It is for this reason that experienced qualified tax professionals should always be consulted when investing offshore. Although this book is designed to give you a firm grounding in order to explore the opportunities offshore, it is essential you take financial advice before you embark on any offshore investment plan. If you already use a professional adviser for your finances, then he or she would be your first port of call. They will be intimate with your financial background and in the best position to advise on what course of action, if any, you should take. In addition, if your adviser does not have the expertise you require, they will be able to recommend an appropriate adviser who does for that particular aspect of your finances.

It is often the case that you will have to use a number of advisers. This is particularly true of expatriates who may well need such eclectic advice to suit a mobile lifestyle. Of course, circumstances must dictate whether you need different advisers. But perhaps the strongest argument is that one advisory firm cannot be au fait with every domestic tax law. For example, Spain doesn't recognise the trust concept but does accept the use of life assurance policies for tax planning.

The basics of choosing a good financial adviser are of course even more important when assessing opportunities offshore.

– Check that the adviser is registered and authorised to give you the appropriate advice.

– Is your adviser charging you a fee or taking a commission on products suggested?

WHAT TO LOOK FOR IN A FINANCIAL ADVISER AND THE COSTS INVOLVED

Major factors include whether your chosen adviser is an offshore specialist adviser; whether it has international links with overseas advisers; or whether it has offices based overseas. For example, Sedgwick Financial Services in London has a division specifically catering for British expatriates, as well as 70 offices around the world. And accountants are often a useful source of multinational advice. Arthur Andersen, for instance, is part of a network operating in 67 countries and can advise expatriates on areas like taxation, inheritance, and strategic retirement and investment planning. A 'lead' adviser will be appointed where input comes from different offices.

Arthur Andersen does not take on day-to-day investment management, preferring to organise a 'beauty parade' of external fund managers or stockbrokers. But to get access to this level of advice from an international accountancy practice, an expatriate would need to be pretty wealthy. As a very rough guide, large accountancy practices are likely to charge £150 or more an hour depending upon the complexity of the advice. Ongoing fees tend to be lower once the initial work is done. Many accountants may be prepared to have an initial free consultation and then quote fees.

Some advisers such as W T Fry in the UK specialise in UK tax planning. The company has trust and taxation divisions giving it an edge on advising returning Britons. Where expertise is lacking, a good adviser should have plenty of reputable contacts acting as a liaison point for expatriates needing pre-return tax planning as well as handing over some or all of the investment aspects of portfolio management. These contacts become crucial if, in the not uncommon situation, an expatriate decides not to return to his home country but retire instead to a different country altogether. Blackstone Franks International is another UK based adviser whose specialist area is in expatriates in southern Europe, particularly Spain, France and Portugal.

Professional qualifications can be a further sign of how committed and experienced an adviser is. Robert Noble-Warren, managing director of UK advisers Murray Noble specialises in international clients, and in particular UK/US situations. He is the only member of the Institute of Financial Planning (IFP) who has dual qualifications in both the UK and the US.

Aside from expertise and qualifications, the ability to communicate with your adviser is all-important, particularly for far-flung expatriates. The ability to communicate speedily is a priority, particularly when world stockmarkets are displaying volatility. Many advisers specialising in expatriate advice have now opened offices in areas of the world where they have clients. UK-based WT Fry recently established a presence in Hong Kong. Another UK adviser, Towry Law International, has advisers stationed permanently in Bahrain, Hong Kong and Switzerland as well as the UK.

Since 1990, the UK-based Institute of Financial Planning (IFP) has been busy drawing together an international network of firms under the auspices of the International Certified Financial Planning Council. The trademark qualification is the Certified Financial Planner (CFP). It is now held by more than 40,000 individuals and is formally recognised in six countries through relevant standard setting organisation.

The founding six organisations are:

–   New Zealand: The Association of Investment Advisers and Financial Planners

– US: Certified Financial Planner Board of Standards

– Australia: The Financial Planning Association

– Canada: Financial Planner Standards Council

– UK: Institute of Financial Planning

Through the Institute, advisers with an expatriate client can contact a specialist in a growing number of countries.

WHAT IF THINGS GO WRONG?

If you invest offshore through an adviser you have more chance of recourse if things go wrong. The complaints procedure, amount of compensation and what you can be compensated for differs from country to country. In Europe, there are guidelines for a minimum level of compensation for faulty financial advice. Some European countries, such as the UK, offer a higher level of compensation than the minimum recommendation. Australia also operates a compensation scheme covering advice on investments.

However, most forms of compensation only cover advice where the business was conducted in the country. So, for example, you will only be covered by the UK compensation scheme if the advice, which led you to an investment offshore, came from an UK-based adviser. This makes expatriates or emigrants particularly vulnerable if things go wrong.

There have been many incidences of people who have been advised to find another country of residence to avoid UK capital gains tax when selling their UK business. They jaunt off to Spain and end up facing the more punitive Spanish equivalent.

In terms of expatriates, it is worthwhile remembering that while the regulatory situation covering advisers operating internationally has improved, commission-driven salesmen are still out there and operating where expatriate enclaves exist. You have a better chance of good advice if you stick with advisers who are covered by their home country regulatory regime and employ salaried salesmen.

In addition, it is important to check out what help an offshore centre can give you. The main procedures if you have a complaint against a group are listed overleaf:

Name of Centre	Investor Protection	Contact details
Bahamas	No investor compensation scheme	Bahamas Chamber of Commerce, PO Box N-532, Nassau, Bahamas. Tel: 809 322 2145 Fax: 809 322 5553
Barbados	No investor compensation scheme	Ministry of Finance and Economic Affiars, Govemmetn of Barbados, Treasury Building, 6th Floor Bridgetown, Barbados.
Bermuda	No investor compensation scheme	Ministry of Finance, Government Administration Building, 30 Parliament Street, Hamilton, HM12, Bermuda
British Virgin Islands	No investor compensation scheme. Considering legislation regulating financial service intermediaries which may include such provisions.	Financial Services Department, Government of the BVI, Road Town, Tortola, BVI. Tel: 809 494 4190. Fax: 809 494 5016
Cayman Islands	Only Preferred Payments Scheme for depositors in Category A banks.	Cayman Islands Government Information Services, George Town, Grand Cayman. Tel: 345 949 8092. Fax: 345 949 5936
Cook Islands	No investor compensation scheme	The Secretary, Cook Islands Monetary Board, PO Box 594 Avarua, Cook Islands, Tel: 00682 20 798
Cyprus	No investor compensation scheme	Central Bank of Cyprus, International Division, PO Box 5529, Nicosia, Cyprus
Dublin	In the process of implementing the EU Investor Compensation Directive	The Industrial Development Authority (IDA), IDA Ireland, Wilton Place, Dublin 2. Tel: (01) 6034000. Fax: (01) 671561
Gibraltar	Will implement the EU Investor Compensation Directive	Finance Centre Development Director, Department of Trade and Industry, Europort, Gibraltar. Tel: (350) 74804. Fax: (350) 71406
Guernsey	Yes for investors in Class A schemes. Up to a total of £60,000 per investor	Guernsey Financial Services Commission, Valley House, Hirzel Street, St. Peter Port, Guernsey, Channel Islands. Tel: 44 (0)1481 712706. Fax: 22 (0) 1481 712010
Isle of Man	Yes	Commercial Development Division, Treasury, Dept. 500, Government Offices, Douglas, Isle of Man. Tel: 44 (0)1624 685755. Fax: 44 (0) 1624 685747
Jersey	No	Financial Services Commission, Cyril Le Marquand House, PO Box 267, The Parade, St Helier, Jersey JE4 8TZ , Channel Islands. Tel: 44(0)1534 603600. Fax: 44(0)1534 89155
Liechtenstein	No	Liechtenstein Government, Regierungskanzle, FL-9490 Vaduz, Liechtenstein
Luxembourg	No investor compensation	Institut Monetaire Luxembourgeoise (IML), 63 Avenue de la Liberte, 22983, Luxembourg. Tel: 352 40 2929250
Netherlands Antilles	No investor compensation	Bank van de Nederlandse Antillen (Central Bank), Breederstraat1, Willemstad, Curacao, NA. Tel: 599 9 327000. Fax: 599 9 327590
Switzerland	No investor compensation	Federal Banking Commission, Marktgasse 37, 3001 Bern. Switzerland Tel: 41 31 322 69 11
Turks & Caicos	No investor compensation	Superintendent of Financial Services Commission, Grand Turk, Turks & Caicos Islands, British West Indies. Tel: 649 946 2937. Fax: 649 946 2557

Investor protection

CHAPTER 5

TRUSTS

WHAT IS AN OFFSHORE TRUST?

The concept of a trust is a difficult one for many investors to understand and accept. This is because the fundamental basis of a trust means you no longer own or have control of the assets you place in the trust. Such assets may include real estate, stocks, bonds, works of art and intellectual property like patents and copyrights. It is usually created in an English common-law jurisdiction because the law governing trusts has developed in English court cases over the centuries. This case law became a part of the English common law, not only in the United Kingdom, but also in most of the former and present British territories of Canada, United States, Channel Islands, Bahamas, and Cayman Islands. Trusts where the strategy covers more than one legal jurisdiction can therefore be administered effectively in the many offshore low tax areas that have English-based law.

Why anyone would want to separate themselves from their wealth is the stumbling factor for most investors. Some of the main reasons for this follow. The concept is doubly difficult for investors of civil law jurisdictions to understand such as the Middle East or France. Civil law systems do not generally accommodate the distinction between legal ownership and beneficial ownership on which the existence of a trust depends, and therefore do not easily recognise the concept of a trust.

This appears to be changing, albeit slowly. While civil law courts have been willing in the past to recognise trusts formed by individuals not connected with their jurisdiction, the increased global use of trusts by individuals, institutions and businesses means that civil law jurisdictions, such as Germany and France, are beginning to recognise trusts formed by their own nationals.

The fact that trusts have been part of common law jurisdictions since medieval times means there are many legal precedents which guarantee trusts fulfil what they set out to do. These legal precedents ensure that the people (trustees) who hold the trust assets on behalf of the people (beneficiaries) who will benefit from the trust do so according to your wishes when you (the settlor) established the trust. Trustees are governed by a strict ethical code that makes sure they always act in the best wishes of the trust letter (the rules governing how the trust is administered). It is heartening to know that there is a growing trend for trustees to come under scrutiny if the beneficiaries feel that the trustees are being unreasonable. Offshore centres are increasingly introducing guidance rules on this that gives recourse to beneficiaries who have a valid complaint against the trustees.

Establishing an offshore trust has much the same qualities as other offshore investments in that being placed (sited) in a low or no tax regime potentially gives a performance kick on the underlying investments held in the trust. It also gives additional financial planning flexibility, particularly if at some time in the future you are likely to become resident of a civil law jurisdiction where the distribution of your assets on death are subject to strict guidelines. France's forced heirship rules are an example of this.

The avoidance of forced heirship is a key use of a trust in civil law jurisdictions where complete testamentary freedom is not available as a matter of public policy. Forced heirship rules should not be confused with intestacy rules that apply where the deceased has not left a valid will. Forced heirship rules give a fixed entitlement or proportion of a deceased's estate to certain of the deceased's close relatives. So if, for example, the deceased had made capital gifts during his or her life time, these may well be "clawed back" and distributions out of the estate on death adjusted accordingly or, if necessary, actions taken to recover part or the whole of the gift.

Since forced heirship rules are a matter of public policy, any attempt to override them may be illegal. Following Roman and Napoleonic codes, civil law in continental Europe tends to require that children have the right to a fixed amount of their parents' estate. In certain Arab jurisdictions, the principle of combating forced heirship can apply in cases where sisters are sanctioned to inherit less than brothers do.

Expatriates resident in European civil law jurisdictions may think that the forced heirship rules do not apply to them because they are not nationals of that jurisdiction. But if you become permanently resident and there is a chance you could be deemed to have domicile status, you could be subject to forced heirship rules. By creating a trust before becoming a permanent resident then any property held in trust should fall outside the heirship rules.

In theory, establishing a trust requires a substantial amount of wealth to justify the effort and expense involved. However, increasingly simpler trust versions are being used in conjunction with offshore investment products in order to maximise tax planning opportunities. For example, offshore bonds are often written in trust to mitigate inheritance tax. In some instances, the popularity of such trust use has made them a victim of their own success and tax authorities are generally keen to close such blatant loopholes. This was recently witnessed in the UK when action was taken to put an end to the 'dead settlor provision'. The dead settlor provision was the ability to arrange for your heirs to avoid paying tax on the proceeds of insurance bonds by putting them in trust. When the settlor died, there was no one the Revenue could tax.

It is crucial, therefore, to gain a good feeling for how the jurisdiction in which you are resident views the different types of trust use as well as choosing a sound jurisdiction favourable to the type of trust you wish to establish. The following information will give you a good grounding for what to look out for.

Name of Centre	Political stability	English Common Law	Domicile Migration	Trusts	Asset Protection Trusts	Trust perpetuity
Bahamas	Excellent	Yes	Yes	Yes	Yes	Life+ 21 years
Barbados	Very Good	Yes	Yes	Yes	No	Life +21 years
Bermuda	Excellent	Yes	Yes	Yes	Yes	100 years
British Virgin Islands	Excellent	Yes	Yes			100 years + 'wait and see' rule
Cayman Islands	Excellent	Yes	Yes	Yes	No	150 years + 'wait and see' rule
Cook Islands	Very Good	Yes	Yes	Yes	Yes	Unlimited
Cyprus	Good	Yes	Doubtful	Yes	Yes	100 years
Dublin	Excellent	Yes	Doubtful	Yes	No	n/a
Gibraltar	Very Good	Yes	Doubtful			100 years + 'wait and see' rule
Guernsey	Excellent	Yes (for trusts)	Doubtful	Yes	Yes	100 years
Isle of Man	Excellent	Yes	Doubtful	Yes	No	80 years or Life + 21 years
Jersey	Excellent	Yes (for trusts)	Yes	Yes	No	100 years
Liechtenstein	Excellent	Yes (for trusts)	Yes	Yes	Yes	Unlimited
Luxembourg	Excellent	No (civil law)	Probable	No	No	n/a
Netherlands Antilles	Excellent	No (Dutch civil)	Yes	No	No	n/a
Switzerland	Excellent	No (civil law)	Yes	No	No	n/a
Turks & Caicos	Excellent	Yes	Yes	Yes	Yes	Unlimited

Trusts

THE DIFFERENCE BETWEEN COMMON LAW AND CIVIL LAW

The legal concept of the trust is inherent in all common law jurisdictions. The basic features of trust law are essentially the same in all common law jurisdictions. Basic features of trust law recognise the roles and relationship between the three basic parties to the trust. The settlor, who creates the trust; the trustee, who holds legal title to the trust assets; and the beneficiary, for whose benefit those assets are held and administered. The lack of an inherent trust concept in civil legal systems, even though an enabling trust law may have been enacted by statute, injects an element of uncertainty into this type of planning. It is important, therefore, to select a trust jurisdiction whose legal system is common law-based.

TRUST DEFINITIONS

One of the most popular vehicles for protecting assets, reducing taxes and controlling wealth is through an offshore trust. A trust is not regarded in law as a legal "person" such as an individual or company. It is a set of obligations which binds the holder of property to hold the benefit of that property for other persons and not for the holder's own beneficial interest. The trust is a concept of the English common law system and had its origins in the English property law of mediaeval times.

Albany Life International, a life company based in the Isle of Man and specialising in offshore life-linked investment products, gives the following definition of a trust and how it works:

A trust is a way of arranging property for the benefit of people (beneficiaries) without giving them full control over it. This may be done for reasons of convenience. For example, if the beneficiaries are to be minor children who could not appreciate or deal with the property themselves. Or to mitigate some form of taxation.

The essence of the arrangement is that the trustees have the legal ownership of the property, but cannot use it as their own personal property: they have to use it for the benefit of the beneficiaries. In every trust there is, therefore, this division of ownership. The trustees will be the legal owners and, for example, would be entitled to claim against the life office if the trust property included a life policy. The beneficiaries have the equitable or beneficial ownership that means that although they cannot claim against the life office they can claim against the trustee and beneficiary.

How a Trust Works

Every trust has a settlor, trustees and beneficiaries. Trusts written under some legal systems often also incorporate a protector.

The Settlor: The person who sets up the trust is called the 'settlor'. The settlor is the original owner of the property being put into the trust and thus has to transfer legal ownership of it to the trustees. This is usually done by a deed of trust (sometimes called a settlement) but it can be done in other ways.

The Beneficiaries: The beneficiaries of a trust are the equitable or beneficial owners of the property and everything that takes place must be for their benefit. An individual can be a sole beneficiary or one of several. Joint beneficiaries will take benefits in equal shares unless the trust states otherwise. A beneficiary may be named or defined by description or class. Often beneficiaries are defined as a class for extra flexibility. For example, the class of beneficiaries may be the children of the settlor. Therefore, if the settlor has further children subsequent to the formation of the trust, they can still become a beneficiary. Beneficiaries cannot control the trustees, but they can insist that the trustees act in accordance with the trust deed and general law. Under a trust where the beneficiaries are ascertained (as in an absolute trusts below) and there is no possibility of further beneficiaries, they can direct the trustees to distribute the property to them. If a beneficiary is a minor then no demand can be made, but the trustees have powers to advance income for the beneficiary's education and maintenance.

The nature of their interest in the trust property can take several forms:

Absolute Interest: the beneficiary has a full equitable interest which cannot be taken away. Provided he or she has attained 18 years of age and is of sound mind, they may wind up the trust in respect of the property by simply demanding it from the trustees.

Life Interest: a beneficiary has a life interest when entitled to the income from the trust property for life but cannot receive capital. Such a beneficiary is called a life tenant. After a life tenant dies, the property passes to the next beneficiaries who are called remaindermen. The remaindermen only get a full interest after the death of a life tenant; until then their interest is known as a reversionary interest.

Contingent Interest: a beneficiary's interest is dependent upon a particular event, in other words the contingency, which may never happen. For example, a beneficiary may only attain an interest on surviving to a specified age.

Potential Interest: The beneficiary is defined in the trust deed as only achieving a vested interest in the trust property when the trustees exercise a discretionary power to appoint such a vested interest in the beneficiary's favour.

The Trustees: The normal number of trustees is usually between two and five. It is theoretically possible to have a single trustee, however this is usually inadvisable. For example, if the trustee was to die, then probate would be required.

Duties of Trustees: The job of the trustees is to hold the trust property and administer it for the benefit of the beneficiaries, as directed by the trust provisions or deeds. Trustees should hold the title document to any trust assets. For example, policy documents for life policies as well as ensuring that they are registered as the legal owners on any relevant registry, e.g. land registry or register of shareholders of a company. They must make sure that everything they do with the trust assets is done for the benefit of the beneficiaries and is authorised by the terms of the trust. The trust deed will usually give the trustees specific powers to deal with the trust property. If any cash comes into the trust, the trustees have a duty to invest it unless it is being paid to a beneficiary straight away. Any investment must be authorised by the trust deed and modern trusts contain wide powers of investment. However, if the deed has no specific investment powers (many older trusts do not) then the investment vehicles the trust can hold are limited. When a trustee is carrying out his duties, he must use utmost diligence to avoid any loss and will be liable to the beneficiaries if he acts in breach of his duties. The trustees must keep property accounts of all trust property. The beneficiaries are entitled to see the accounts on demand and can also require any information about the dealings of the trust that is reasonable. Trustees are liable to the beneficiaries for any loss caused by their dispute unless they can produce evidence to show that they acted in good faith. Activities which should be recorded include: investment decisions, distributions from the trust fund to beneficiaries (including loans), changes to the class of beneficiary, re-appointment of beneficiaries' interests, and appointment and removal of trustees. Such matters as changes of trustees or exercise of discretionary owners should be formally recorded by deed. Trustees cannot charge for their services unless they are permitted to by a trustee-charging clause in the trust deed. This is normal practice for corporate trustees (see below).

Appointment of Trustees: Trustees are usually appointed at outset by the deed setting up the trust. Individuals should be over 18 and of sound mind. Trust corporates can also be trustees and have several advantages over individuals. Firstly, trust corporations can be a sole trustee as they are a legal entity that cannot die. Secondly, they can call upon legal, investment and taxation expertise not normally available to individuals. Obviously, corporate trustees will charge for their services. If an individual trustee dies it may be necessary to appoint a replacement. Therefore, many trust deeds name an individual as appointor who is entitled to appoint new trustees. If the deed has no provision, the power to appoint new trustees is possessed by the surviving trustee(s) or the legal representatives of the last trustee to survive. A trust will never want for the lack of a trustee and there are statutory provisions to rectify a trust with no trustees.

Retirement of Trustees: If a trustee wishes to retire, he or she can usually be replaced. A trustee can normally retire without being replaced if at least two trustees remain (or a trust corporation). The co-trustees and appointor must consent in the deed.

Death of Trustees: If a trustee dies, there is generally provision within the overseeing trust acts to allow for a replacement. However, if this is not done, the powers of a deceased trustee can be acted on by the surviving trustee(s). If a sole or last surviving trustee dies, then his or her legal representative can act as a trustee.

The Protector: The protector is an additional party found in many trusts where it is felt desirable that there should be a guardian over the trustees' activities. Even though trusts written under English law can have a protector according to statute, a protector is not used on such trusts today as the powers of the protector have been substantially eroded by the courts. However, this is not the case in other legal jurisdictions where protectors still have effective powers. The protector's powers are defined in the trust deed, the terms of which would usually require the trustees seeking the protector's agreement before exercising their trustee powers. Examples of the protector's powers are to veto over any distributions of income or capital to beneficiaries proposed by the trustees, veto over loans to beneficiaries proposed by the trustees, to be able to dismiss trustees and appoint new or additional trustees, and to nominate a successor protector.

DIFFERENT TYPES OF TRUSTS

There are two main types of trust:

1) Fixed Interest

Absolute: An absolute trust has beneficiaries with an absolute interest and the sole duty of the trustees is to hold the property for the beneficiary and transfer it to the beneficiary when required. If the beneficiary is a minor, the trustee has to investment money until the child reaches majority and then transfer it to him or her.

Life Interest: The trust involves a life interest. For example, a trust can be worded to say 'my wife for life and thereafter to my children (A, B and C) in equal shares'. The wife is the life tenant and on her husband's death is entitled to income from the trust for her life. On her death the trustees split the property three ways and distribute to each beneficiary. The children in this example are the remaindermen and are said to possess the reversionary interest, at least until the death of the life tenant.

2) Flexible Trusts

Power of Appointment: This type of trust, together with the discretionary trust described below, differs from the fixed trusts in that the trustees have the power to vary the beneficiaries. On the power of appointment trust, the beneficiaries can be varied within specified limits. It can thus be altered to cater for changing circumstances.

Discretionary: The distributions from a discretionary trust are solely at the discretion of the trustee. In addition to the settlement giving either fixed interests or flexible trusts, it is also possible in certain jurisdictions to set up charitable trusts, purpose trusts and asset protection trusts.

Charitable Trusts: will be devoted to charitable purposes and will usually provide that the trust corpus is held on trust for a particular charity or charities or for a specific charitable purpose. Charitable for this purpose under an old statute of Elizabeth I is defined as being for the relief of poverty, the advancement of education or religion or other purposes similarly beneficial to the public at large.

Purpose Trusts: are trusts that are established for a purpose or purposes that are specific, reasonable and possible and are not immoral or contrary to public policy or unlawful. A trust for a purpose under the Act means a trust "other than a trust that is for the benefit of a particular person whether or not immediately ascertainable or some aggregate of persons ascertained by reference to some personal relationship."

Asset Protection Trusts (APTs): Although all trusts protect assets, the proliferation of multimillion-dollar lawsuits in US courts has led to the creation of a particularly noteworthy type of trust with the misleading title of Asset Protection Trust (APT). Most APTs are designed to defeat the claims of present or future creditors of the trust's grantor. Whereas generally the law will not allow trusts to defeat or defraud current or known pending creditors, some offshore jurisdictions have made APTs attractive by shortening the time within which an action can be brought by creditors to set aside the transfer of assets to the trust as a fraudulent conveyance. APTs have become popular vehicles for doctors, dentists and other professionals who practice without the benefit of limited liability protection. Settlements or damages awarded by juries, particularly in US courts, are often in excess of the insurance coverage that these professionals are able to arrange in a highly litigious society. Also, stringent divorce laws and the forced heirship laws of certain countries can bring about the division of assets - which may have been built up over a long period of time, perhaps even generations. Although APTs could prove effective against unknown future claimants, they have not been fully tested in the courts. The most commonly used offshore locations for APTs are the Cook Islands, the Cayman Islands, the Bahamas, Bermuda, Gibraltar, and Cyprus.

REASONS FOR ESTABLISHING A TRUST

Controlling Family Assets

Often the beneficiaries of a trust will be minor children who could not appreciate or deal with property. Therefore, the settlor may wish to restrict the beneficiaries' access to the trust property until it is thought appropriate that the property should be distributed. Trusts initially and commonly still are used to protect the wealth of individuals and family groups. And they can do it discreetly. In many jurisdictions there is no requirement to register the trust publicly or otherwise disclose the contents of the trust deed, accounts or other sensitive information. This provides anonymity for the parties involved and professional trustees are well aware of the need to maintain confidentiality at all times.

Avoiding Probate Delays

An offshore investment not written in trust may well run into problems if the investor dies, as the heirs will have to prove a Will in the offshore centre in order to claim any proceeds. Writing investment vehicles in trust negates this problem. The uses to which a trust can be put are, of course, as numerous as the ingenuity of trust lawyers can devise. However, it may be useful to summarise some of the principal uses for which trusts have been formed:

Tax Planning: Tax advantages can be gained through the use of trusts by virtue of the separation of legal and beneficial ownership. This can include pre-immigration trusts as well as transfer of assets abroad to avoid inheritance or capital gains taxes, although as discussed above, this element of tax planning using trusts should be taken with great care in the light of growing crackdowns by onshore authorities.

Estate Planning: To get around forced heirship provisions of law in the settlor's domicile or to overcome difficulties caused by community of property regimes. To provide certainty of jurisdiction where problems with domicile might otherwise result. Also, secrecy, particularly in countries where the family does not want its wealth known to avoid kidnapping and similar crimes of extortion.

Business: Pension plan trusts and other deferred compensation schemes as well as similar employee stock option plan trusts.

Protection of assets: This is a particularly controversial use of trusts and not all offshore jurisdictions will allow them. Basically, it is to protect professionals such as doctors and dentists from potential creditors following litigation (See Asset Protection Trusts above). Into this category you can also include pre-marital trust agreements.

These are just a few examples of the more common uses of trusts but is by no means exhaustive. Further examples of the uses of trusts are to avoid confiscation of assets by a hostile regime; to avoid the future imposition of exchange control or other regulations restricting the movement of capital; and to provide a central enduring control of assets for geographically widely scattered beneficiaries and investments.

FACTORS WHEN SELECTING AN OFFSHORE TRUST JURISDICTION

Taxes: It is important to ascertain whether any income or capital taxes are levied on non-resident trusts in your chosen centre. Also whether settlements that are non-resident are free from exchange controls and stamp duties.

Investment Powers: Clarify what is recognised as a valid trust investment. Where authorised trust investment powers are limited, it is possible to include an investment clause permitting the trustees to have much wider ranging investment powers. Obviously, this does not exclude the trustees from carrying out his or her obligations to select investments that are suitable for the trust and its objectives.

For how long can a trust be established? Under the Perpetuities and Accumulations Act 1989 a trust can be established for a specified period of up to 100 years. It is also possible to adopt the old rules permitting a perpetuity period of a life or lives in being plus 21 years (the so called King George clause - as frequently the life or lives in being selected are the descendants of King George V living at the date of the settlement)

Can the trust deed be changed? If the trust is expressed to be revocable by either the settlor or the trustees, the terms of the trust can be changed by revoking the trust and resettling the trust corpus on the terms of a new trust deed incorporating the required changes.

If the trust is expressed to be irrevocable it may be possible (if all adult beneficiaries agree) to make application to the courts for approval of any variation of the provisions of the settlement. However, as the usual form of settlement is discretionary, this may be unnecessary, as there is usually sufficient flexibility in the provisions of the settlement itself.

Flight Clause: It is usual to include in the settlement a power to move the trust to another jurisdiction (the so called "flee clause"). Such a provision is commonly included in an offshore trust to address various situations, not least of which is civil unrest, or an unfavourable change in the law or political climate of the domicile jurisdiction. The power is usually given to the Protector or the Trustees and to apply the laws of that jurisdiction henceforth as the governing or administrative law of the settlement. Power is usually given under the settlement at that time to make any necessary amendments to the provisions of the settlement to conform the settlement to the laws of the new jurisdiction.

Geographical accessibility: In the eventuality you may need to visit the jurisdiction, ensure that the location has easy access by air to and from most major cities.

Political stability: It is important to ensure that the jurisdiction is politically stable to ensure that the appropriate trust laws are not amended causing a detrimental effect on the original trust deed.

Sophisticated infrastructure: Stick to jurisdictions with experienced lawyers, accountants, trust professionals and advisers to ensure the smooth running of the trust. This includes first class communication facilities such as telephone and other electronic forms of communication.

Retroactive Application of Law: Sometimes it may be appropriate to redomicile a client's existing trust to an offshore jurisdiction (or from one such jurisdiction to another). Under such circumstances, statutory certainty regarding which jurisdiction's law will apply in litigation may become crucial.

Protecting your Trust
When establishing a trust, protection can be enhanced through careful drafting.

The following is a brief description of certain provisions that may be included in a trust deed to enhance protection:

Trust Protector Clause: The Protector is featured on page 63. However, as it is not a mandatory feature of a trust, it is worth reiterating the benefits of appointing a protector in terms of enhancing trust protection. Including a trust protector clause enables a designated person or persons, other than the trustee, to exercise certain powers with respect to the trust. Legislative recognition of the trust protector concept in offshore jurisdictions varies from statutes providing the trust protector with specified powers, to statutes that merely recognise or define a trust protector. Examples of common trust protector powers include: removal and replacement of trustees, veto of discretionary actions of trustees, consent to trust amendments proposed by trustees.

Spendthrift Clause: A spendthrift clause is a restraint on the voluntary or involuntary alienation of a beneficiary's interest in a trust. Such clauses can provide for a forfeiture of a beneficiary's trust interest upon an attempt by the beneficiary to transfer it, or by his creditors to reach it.

The Hague Convention

The Hague Convention was designed to provide an international legal framework that would allow trusts increasingly to cover more countries. For the purposes of the Hague Convention, the term "trust" refers to the legal relationship created - inter-vivos (between living individuals) or on death - by a person, the settlor, when assets have been placed under the control of a trustee for the benefit of a beneficiary or for a specified purpose. It is now widely accepted that the Hague Convention provides the universal definition of the trust concept.

CHAPTER 6

COMPANY FORMATION

WHAT IS AN OFFSHORE COMPANY?

The two principal reasons for using an offshore company are the tax benefits it can confer and confidentiality. In terms of confidentiality, an offshore company has a separate legal identity so that while an individual may beneficially own the company, it will have its own name and that name can be completely disassociated from the beneficial owner.

Not only is the beneficial owner's identity safeguarded but also the financial obligations of the shareholder are guaranteed because offshore companies carry limited liability status. Establishing an offshore company within a well-structured offshore jurisdiction is a smoothly organised affair. Offshore company providers will first arrange permission from the local authorities for a company name. Many offshore centres now boast the ability to establish a company within 24 hours. Necessary paperwork such as the Memorandum and Articles of Association generally come in standard form and give the company wide-ranging powers.

Offshore companies require two subscribers who will agree to become shareholders. For those seeking further guarantees of anonymity, it is possible to buy the services of nominee shareholders. Offshore companies can be easily dismantled and assets distributed back to the owners. It is quite common to establish a trust in conjunction with a company.

REASONS FOR ESTABLISHING A OFFSHORE COMPANY

The uses of offshore companies include the ownership of foreign assets and investments (listed and unlisted securities, fixed term deposits, intellectual property, real estate, royalty agreements, ships, aircraft, etc.), international trading, and investment management. Also, since most types of companies have continuity of life; its assets are not subject to death duties or estate taxes.

Tax: The tax advantages to be derived from an offshore company will depend on your own personal circumstances. Take the example of inheritance tax mitigation. An offshore company owns £300,000 worth of UK equities, and on the death of the beneficial owner the beneficiaries are bequeathed shares in the company that holds the equities. The crux here is that it is the shares of an offshore company that are bequeathed and these do not necessitate a report being made to the capital taxes office on the death. Obviously the beneficiaries may suffer a tax liability depending on their current residency status. Using an offshore trust to support an offshore company is one way of giving added protection.

Offshore Trading: In many cases, offshore companies are used as a means to facilitate offshore trading. Most offshore centres will stipulate the appointment of a locally-based director. When a company is more actively used, the position of director takes on more weight. Not least this ensures that the company is controlled and managed in a fiscally neutral jurisdiction. The key duty of a director is to manage the company's business. The provision of nominee directors is something people look very closely at before they agree to act. Under company and bankruptcy laws, directors can be personally liable for debts of a company if there has been any fraudulent trading in any way, such as trading when the company is insolvent.

Individual Services: Professional services, including those of authors, consultants, designers, or entertainers may reduce or eliminate tax liability by interposing an offshore employment company. The company's profits may be subject to minimal, if any, taxes that can be reinvested more efficiently in the tax beneficial offshore environs than in one's home country. Payments to the individual "employee" can be structured to limit their tax consequences. Taxation of repatriated funds varies from country to country so consulting a tax advisor is strongly suggested.

Confidentiality: At times it is necessary for an individual or a company to restrict competitors, family members, or predators access to the individual's/company's true financial position. Even the ownership of a company can be disguised. Because such privacy is not available in many countries, utilising a company, trust, and/or establishment/foundation in an offshore site that assures confidentiality is often used.

Intellectual Property: It is not unusual for owners of copyrights, patents, or trademarks to assign the rights of their intellectual property to an offshore company. The company can develop sublicensing agreements worldwide and exploit beneficial international withholding tax rates for dividends, royalties, and/or interest that exist through treaties. Some strategies also involve holding company and subsidiaries.

REGULATION AND COSTS

Up until recently, company formation activities have been fairly unregulated. However, many centres are now in the process of establishing laws governing the regulation and control of this sector. In the meantime, it is important to check out the records of those offering offshore company formation services.

Although it is possible to buy an offshore company off-the-shelf quite cheaply, there are other costs to bear in mind. For example, the total initial set up fee for a Jersey-based offshore company will be around £500 plus a further £150 paid to the local authorities. The annual management fee with all the nominee services, registered office, secretarial, administrative service will probably be around £1,000 a year. The starting price for an Isle of Man-based company is £250, but expect this figure to rise to as much as £1,700 for initial formation and first year costs which can include management fees, directors, government taxes and opening a bank account. Thereafter, the annual running costs are around £1,300. Establishing a Gibraltar-based company will attract similar set-up costs to the Isle of Man. Although for relatively inactive companies, after paying £200 in government duty and anywhere between £250 to £400 for company formation and the same again for annual management, the overall costs look marginally cheaper. Wherever the company is based, further charges must be expected for hiring a nominee director. Fees for nominee directorships are usually done by negotiation depending on the level of responsibility. On the basis that such duties will not be too onerous then expect such a charge to be around £500. For a more active directorship, attending board meetings, more hands-on managing and dealing with the paperwork, expect the fee to be calculated on a time basis.

DIFFERENT TYPES OF COMPANIES AND THEIR USES

There is fierce competition between offshore centres on company formation. Centres are continually updating the laws on company formation to both reflect the changing market and offer a flexible and secure base to encourage offshore company formation. Guernsey in the Channel Islands recently made amendments to Guernsey Company Law relating to crossborder migration. The Ordinance on crossborder migration enables companies incorporated on the register of another jurisdiction to be transferred from that register to the Guernsey register, thereby become a Guernsey incorporated company without the need to reform the company. This is particularly useful if a change of jurisdiction is needed for business purposes, as it avoids the necessity to liquidate assets in order to register the company. The British Virgin Islands (BVI) has been particularly successful in company formation. The vast majority of companies incorporated in the BVI are International Business Companies (IBC). The register of IBCs in the BVI now exceeds 250,000. Outlined below are the most common forms of offshore company. As with any offshore product, centres offer a variety of variations and it is important to check what is most suitable for your own circumstances. A checklist of features you should look out for are:

Exempt Companies: Companies, established in certain jurisdictions, may be exempt from the usual local tax rates to its foreign source income (including capital gains). Generally, an exempt company may not be beneficially owned by a resident of the jurisdiction nor own local real estate, and if it trades or is commercially active in the jurisdiction, its local income will be taxed at the standard rates and may lose its exempt status. There are two types of exempt companies: resident and non-resident. This distinction is important for two reasons. If the beneficial owner or management resides in a high tax country (e.g., UK or US), the tax authorities of these countries may easily establish that the company is really still resident "onshore" and tax the company regardless of its non-resident offshore address. To avoid this dilemma, it would be wise if the directors' meetings and management decisions were made, or appear to be made, in the offshore jurisdiction.

Limited Liability Companies: The Limited Liability Company (LLC) is a form of business entity that has been available in Germany (GmbH), France (SARL), Mexico (Limitada), the US and in various offshore centres (e.g., Bermuda, Cayman Islands, Turks & Caicos Islands) as the Limited Duration Company (LDC). Many of the more recent offshore centre applications of LLCs have been based on the Wyoming model. An LLC is mainly used to reduce exposure to personal liability and is thus similar to an Asset Protection Trust. It also reduces unnecessary double taxation. An LLC has members. It is not a corporation, a partnership or a sole proprietorship, although it is a blend of some of the best characteristics of them all. Key features of a LLC are:

- It has a separate legal existence (legal personality and capacity).

- It must be dissolved within a set period, usually within 30 years of being formed.

- Liability of members is limited to their contributions.

- Unanimous consent is required for the full transfer of an interest.

- It is managed by its members, although they can appoint a manager.

- It is wound up on a change of membership, unless members elect to continue.

- Its profits are treated as the income of its members for tax purposes.

International Business Company (IBC): The International Business Company (IBC) legislation is peculiar to a few offshore centres: Bahamas, Barbados, Belize, British Virgin Islands, Cayman Islands, Jersey, and Turks & Caicos. These quintessential offshore companies must be owned or controlled by non-residents (individual, corporations, etc.) and generally cannot carry on any local business or invest in local property or securities. Advantages include quick formation and low fees, confidentiality of ownership and directors, low or no taxes on foreign profits, and minimum accountability. Essentially, an IBC operates internationally, investing in stocks, bonds, commodities, real estate and so on. They are often used as holding or trading companies, or combined with trusts in complex tax minimising structures. The IBC has enjoyed great popularity with international investors in recent years and the reasons for the upsurge vary. The British Virgin Islands, which has the greatest number of IBCs (more than 250,000), was certainly helped by the fallout of Panamanian offshore companies in the late 1980s and the transfer of Hong Kong in 1997. Some offshore jurisdictions, such as the Bahamas, Cayman Is., and Turks & Caicos, have aided their own cause by recently introducing or updating their IBC legislation to attract more customers. Superficially, IBCs from all jurisdictions may appear alike. In fact, they may even seem indistinguishable from "Exempt" companies. But there are real differences and even the subtle differences can be quite critical when choosing the IBC's domicile. Regardless of jurisdiction, the typical characteristics of an IBC are:

- No, or very low, taxation (capital gains, income, or withholding) or stamp duty on financial instruments.

- Quick formation, if "shelf" companies are not available.

- Very low incorporation and annual fees.

- No or low minimum capital requirement.

- Easy transfer of IBC to or from the jurisdiction.

- No residence requirements for directors or shareholders.

- Minimum 1 director (may be a corporation) and 1 shareholder.

- No public disclosure of directors, officers, or shareholders.

- Bearer shares and registered shares are permitted.

- Limited liability of owners.

- No report of change in shareholders is required.

- No auditing or filing of annual returns with the Government is required.

Because an IBC operates internationally, its domicile typically does not impose foreign exchange controls. It must have a local registered office and there are many firms that offer registered agents, nominees, and professional services. IBCs are permitted to open local bank accounts, hold meetings, retain certain services, prepare and keep books without being considered to be carrying on local business. The IBC must keep proper books and records at its registered office, but often these records are not open to public inspection. It is not permitted to carry on certain types of operations such as banking, insurance, reinsurance, and trusteeship; to invest or own local real property except by holding a lease of property for use as an office; or to provide registered office services to other companies. It is important to note that, at least in The

Bahamas, IBCs may hold securities in other local companies, whether IBCs or regular. Thus, a Bahamian IBC can be a parent company of the above excluded entities, e.g., a bank.

TIPS ON ESTABLISHING AN OFFSHORE COMPANY

When considering establishing an offshore company certain features you should look out for include:

Director and Shareholders: How many are needed. Obviously the more needed, the costlier the formation of a company becomes. Is there any restriction on nationality or residence when it comes to appointing directors or shareholders? Can they be individuals or corporations? Is it a stipulation that shareholder and directors meeting are held in the offshore centre?

Confidentiality: Are there any documents that have to go on public record. If so what are they? Who should be notified if any changes to the company are made? Are bearer shares permitted? Can shares be held in the names of nominees for the beneficial owners?

Tax: Will the offshore company suffer a tax liability in the offshore centre in any form.

Formation: How long does it take for an offshore company to be incorporated? Just as important - how easy is it to dismantle a company structure if needed.

Capital: Is there a minimum capital requirement to establish a company. What types of shares can be issued? For example, can shares be issued with or without par value? Also, can multiple classes of shares be issued, including common shares, preferred shares, limited shares and redeemable shares and shares that allow participation in certain assets only? Can shares be registered or bearer shares and can they be issued in any currency if required?

Migration: Is there the ability to redomicile the trust? What is the procedure?

Protection of Assets: What powers do the directors have to protect the IBCs assets? For example, if it benefits the IBC, its beneficial owners or creditors, do directors have the power to transfer any or all of the IBCs assets to trustees or another legal entity?

FACTORS WHEN SELECTING AN OFFSHORE COMPANY JURISDICTION

The features you should look for when deciding which offshore centre are no different from establishing a trust in that you need a centre that has political stability and low taxes. However, one of the key factors regarding company formation when choosing an offshore centre is that there should be minimum requirements in relation to issuance of shares, designation of directors and shareholders and maintenance of records.

Political and economic stability: This is essential if an investor is to have faith in establishing an offshore vehicle. Key features here will be how long the centre has been designated as an offshore centre. Are there any signs of the local population resenting the financial sector that could lead to political problems in sustaining its offshore financial status? To what extent does the centre rely on the financial sector for its economic success? Is this is to the detriment of other sectors of the economy and will this cause problem in the future?

Legislation: It is important to introduce fresh legislation that, in effect, reflects the up-to-date approach as a financial centre. It also means the investor can respond to an ever-changing business environment.

Taxation: While most centres will not impose capital and income taxes, wealth taxes or estate taxes, companies will be subject to an annual licence fee. This differs from centre to centre but taking the BVI as an example; the annual licence fee for a company will be in the region of US$300.

(For BVI resident individuals, income tax is up to a maximum of 20% and for BVI companies that are not International Business Companies; corporation tax is at a maximum of 15%)

Confidentiality: Are there any forms of disclosure? Is there a requirement to prepare or file accounts?

Expertise: Check there is sufficient infrastructure and enough experience and resources to support the establishment of an offshore company. A good gauge of this will be the size of the centre in terms of company formation. A further sign of a centre's experience will be the speed at which a company can be incorporated (24 hours is realisable for the most experienced centres) and dismantled.

Exchange Controls: It is essential that investors can freely move their assets in and out of the centre. The official currency should be a major currency.

Language: English is generally the chosen language of business. This avoids difficulties in communication and the need to prepare documents in more than one language.

Costs: The level of fees charged in establishing a company is always an important consideration in selecting an offshore centre. Costs may vary considerably, with incorporation fees varying from $500 to $3,500. Minimum annual fees vary from $350 to over $1,000, depending on the jurisdiction and the local maintenance requirements.

Name of Centre	Tax Treaties	Company type	Offshore Revenue tax	Owner Disclosure	Domicile Migration	Shelf Companies	Minimum shareholders	Minimum directors	Corporate directors	Secretary required	Bearer shares	Registered office/agent	Local directors	Directors disclosed	Local meetings
Bahamas	Nil	IBC	Nil	No	Yes	Yes	1	1	Yes	No	Yes	Yes	No	No	No
Barbados	Several	IBC	1-2.5%	No	Yes	Yes	1	1	No	Yes	No	Yes	No	No	No
Bermuda	Nil	Exempt	Nil	Yes	Yes	No	3	1	No	No	Yes	Yes	Yes	Yes	No
British Virgin Islands	2	IBC	Nil	No*	Yes	Yes	1	1	Yes	No	Yes	Yes	No	No	No
Cayman Islands	Nil	Exempt	Nil	No*	Yes	Yes	1	1	Yes	No	Yes	Yes	No	Yes	Yes
Cook Islands	Nil	IBC	Nil	No	Yes	Yes	0	1			Yes	Yes	No	Yes, but not publicly	No
Cyprus	Many	Exempt	4.25%	Yes	Doubtful	Yes	2	1	No	Yes	Yes	Yes	No	Yes	No
Dublin	Yes	Exempt	Nil	Not publicly	Doubtful	Yes	2	2	No	Yes	No	Yes	No	Yes	No
Gibraltar	Nil	Exempt	Nil	Not publicly	Doubtful	Yes	2				No	Yes		Yes, but can be nominees	
Guernsey	UK, Jersey	Exempt	Nil	Yes	Doubtful	No	2	1	Yes	No	Yes	Yes	No	No	No
Isle of Man	UK	IBC, Exempt, Guaranteed/Hybrid Companies, LLC.	Nil	No	Doubtful	Yes	1	1	Yes	Yes	Yes	Yes	No	Yes	No
Jersey	UK, Guernsey	Exempt, IBC	0-2%	Not publicly	Yes	No	2	2	No	Yes	Yes	Yes	Depends	Yes	No
Liechtenstein	Austria	Exempt (Anstalt)	Nil	No	Yes	Yes	1	1	No	Yes	No	Yes	No	No	No
Luxembourg	Many	Holding (SA)	Nil	No	Probable	Perhaps	2	1	Yes	No	Yes	Yes	Yes	Yes	No
Netherlands Antilles	Few	Exempt (NV)	2.4%-3%	No	Yes	Yes	1	3	Yes	No	Yes	Yes	No	Yes	Yes (for shareholders)
Switzerland	Many	Holding (AG)	Low	No	Yes	No	1	1	Yes	No	Yes	Yes	Yes	Yes	No
Turks & Caicos	Nil	Exempt	Nil	No	Yes	Yes	1	1	Yes	Yes	Yes	Yes	No	No	No

CHAPTER 7

BANKING

A Dutch scientist who spends his life travelling around war zones is perhaps not a typical offshore banking client, but it indicates the growing breed of busy professionals who now use offshore banking services.

The above example of a client at the Royal Bank of Scotland International (RBSI) in Jersey highlights the need to use a banking operation specifically geared to dealing with the needs of international customers through world-wide access to cash, multi-currency account facilities and international money transmission. Although not impossible, it would be difficult to get the same level of service through an onshore bank branch. Apart from the level of services, there is also the additional advantage of the tax-free climate offshore centres enjoy. However, with the majority of offshore centres offering banking facilities, the choice can be overwhelming.

Luxembourg, the Cayman Islands, Jersey and Guernsey are among those offshore centres with developed offshore banking systems. The good news is that the majority of onshore banks now have some sort of offshore presence, which means getting access to information should be fairly easy. For example, 47 of the 50 largest banks in the world operate in some form from the Grand Cayman.

As many banks will have a presence in several offshore centres, the choice of centre will depend on your current domestic banking arrangements and time zone.

Confidentiality of banking arrangements is a further consideration. Offshore centres have come under attack from both the US and Europe to enact specific legislation enabling government authorities to scrutinise banking records in an effort to stamp out international money-laundering operations.

Many centres, including most recently the Cayman Islands, have now adopted anti-money-laundering legislation, which means they will co-operate with international authorities on narcotics-related crime. Far from being the death knell for offshore centres, co-operation on criminal activities is seen as a way for an offshore centre to enhance its reputation as a legitimate base for offshore financial operations. And while such mutual legal assistance treaties are seen as an effective way of keeping organised crime at bay, legislation does not include banks breaking the confidentiality code on tax offences.

The official line is that offshore banks are not out to help customers evade the tax authorities but at the same they are not here to help the Inland Revenue find out about people. The only time a bank account would become accessible to anyone but the client would be through a court order made by local courts under the money-laundering act, otherwise a client's account remains confidential. Both Jersey and the Caymans operate severe penalties for banks breaking the code of client confidentiality. In the Caymans, for example, this is dealt with through the Confidential Relationships (Preservation) Law.

That said, authorities worldwide continue to put pressure on offshore centres to co-operate on exchange of tax information. Switzerland's banking secrecy has already been pierced on crimes related to tax fraud. In order to stave off unwelcome advances from tax authorities, offshore banks are installing their own mechanisms to guard against unwelcome customers. Reputable offshore banks will only open an account upon receipt of a reference from the depositor's home banker and will not accept large sums in cash. Picture signature identification and a statement on the source of funds are other regulations banks now have to abide by.

Within these guidelines, opening an offshore account is a simple enough procedure. Generally, this can be done through the bank's worldwide network. For example, RBSI clients can open an account through one of the branches dotted around the world such as the UK, Jersey, Guernsey, Isle of Man, Gibraltar, Singapore, Hong Kong and Nassau.

The onset of telephone banking and the Internet has made the process even easier, particularly for clients in the Far East who may use a European offshore banking centre. With no time zone restrictions and new technological advances through the internet set to make the process of opening an offshore bank account even easier, choice is now likely to focus on quality of services offered, minimum investment levels and charges. These vary considerably. For example, Cayman-based Altajir Bank offers a popular call account accepting deposits of as little as US$1000.

RBSI's Jersey banking services start with a minimum deposit of US$5000 which gives access to a range of currencies. An integral part of RBSI's services is access to investment and private banking services. This service is particularly geared to the growing number of financially aware clients who want a service that is halfway between advisory and discretionary. Access to RBSI's private banking services requires a minimum of US$250,000. On the investment side US$100,000 minimum investment will get you a personal investment manager and access to a choice of funds from outside fund managers. Charges are negotiable but are in the region of 0.25% per annum management fee and fund charges of between 1-2%.

Offshore banking also caters for the more financially sophisticated client. For clients with a portfolio of shares, Guinness Mahon Bank in Guernsey offers a collaterised lending service to gear further investment. This type of service has increased in popularity when returns from stockmarkets have been good. Basically, the client gets the upside in the increase in value of the portfolio less the costs. The maximum loan will vary depending on the portfolio held. For example, Guinness Mahon will lend up to 80% of the value of a good gilt fund but only 40-50% for an emerging market fund.

Banking in different currencies is now a major feature of offshore banking. All the major currencies are normally featured as are European, Australasian, South African and Asian currencies. However, the Middle Eastern and Eastern European currencies are increasingly on offer. The currency and chequeing facilities offered by offshore banks allow clients to handle their funds, for the most part, without charges. Access is also provided to clients' funds by the provision of card services and their settlement. Larger sums may be placed on fixed-term or notice deposits in many of these currencies for one month to one year to earn potentially high rates of interest.

The benefit of the multi-currency approach is that it can be tailored to fit the client's requirements.

Name of Centre	Political stability	Offshore Banking	Banking secrecy Law
Bahamas	Excellent	Yes	Yes
Barbados	Very Good	Yes	Yes
Bermuda	Excellent	No	No
British Virgin Islands	Excellent	Yes	No
Cayman Islands	Excellent	Yes	Yes
Cook Islands	Very Good	Yes	Yes
Cyprus	Good	Yes	Yes
Dublin	Excellent	No	No
Gibraltar	Very Good	Yes	No
Guernsey	Excellent	Yes	No
Isle of Man	Excellent	Yes	Strict Common Law
Jersey	Excellent	Yes	No
Liechtenstein	Excellent	No	Yes
Luxembourg	Excellent	No	Yes
Netherlands Antilles	Excellent	Yes	Yes
Switzerland	Excellent	No	Yes
Turks & Caicos	Excellent	Yes	Yes

Offshore banking

DEPOSIT ACCOUNTS

The main advantage of offshore deposit accounts is to have an accessible lump sum of money that attracts a decent rate of interest and grows tax-free. A deposit account is a useful short term home for ongoing financial commitments such as mortgage repayments and school fees. All offshore centres will offer deposit accounts as part of a general banking package. Centres such as Jersey, Guernsey, Isle of Man and Gibraltar that have strong links with the UK expatriate market also offer the option of using the services of a number of UK building societies establishing a presence there.

There are several differences between what the banks and building societies offer. Although many of the societies now offer US dollar denominated accounts, in general banks will offer a wider choice of currency accounts. On the other hand, their specialisation in deposit-taking tends to allow building societies to offer higher interest rates.

There are two types of interest rates: variable and fixed. Variable rates are subject to changes at any time. For example, if it is a sterling based account then the interest rate offered will move broadly in line with UK base rates. Fixed rates, sometimes known as guaranteed, are fixed for a defined period of time. The interest rate offered will depend on the rates available on international money markets. The second is the prevailing view on how interest rates will move over the same period of time as the account period.

If you plan to lock your money away in a fixed rate account then you are likely to receive a higher rate of interest than an instant access account. Obviously if you are likely to need access to your money at short notice then it may be better to go for the lower rate offered by the instant access, as although you generally will be able to access your cash in a fixed rate account you are likely to suffer a financial penalty. Typical notice periods are 30, 60, 90 and 180 days. If the notice is not duly given, penalties on withdrawal are incurred which could negate all the money earned from the improved rate.

MULTICURRENCY ACCOUNTS

Multicurrency accounts are suitable for investors who are, say, paid in one currency but have financial commitments in one or more other currencies. Unlike separate currency accounts, which allow one account for dollars and another for deutschmarks, a multicurrency account houses various currencies in a single account.

The facility allows a depositor to hold anything from a few currencies to dozens - ranging from dollars to yen. An account can be initiated with one set of documentation for the single account that will house all the others. Regular statements will be issued, usually quarterly, for each separate currency under the same account number.

In fact, multicurrency accounts offer similar facilities to standard accounts. For example, as well as facilitating money transactions, multicurrency accounts earn interest and provide cheque books and credit cards. Clients may wish to keep virtually all their funds in Sterling, if this represents the currency of the majority of their assets and liabilities. They may, however, wish to use a US$ denominated card for overseas personal spending in a number of countries. The card can either be automatically settled out of the Sterling balances or out of part of their funds maintained in US$, so that the currency exchange is effected at the client's preferred timing.

These are standard and prudent methods of operating a multi-currency account. However, a client can be more speculative although currency speculation invariably involves a large risk factor. The borrowing of a low interest cost currency such as Yen and Swiss Francs to fund investments which are expected to return an overall gain higher than the interest cost of such borrowing is beneficial, provided the currency remains at least stable over the period of borrowing. For example, ¥10m (Japanese yen) borrowed at 1% interest might then be invested in Indonesian rupees with, say, a 10% return. This can be done easily with generally no conversion charges. Strengthening of the borrowing currency, however, can easily damage severely the result of the whole package. The cost of dealing in currencies, in terms of maintaining funds in more than one currency, is not necessarily expensive. Foreign exchange spreads on buying and selling currencies tend to widen as the amount involved decreases, and much depends on the state of the market and the currencies in question. For many banks the spread represents the entire cost to the client; there are no other percentage costs of the transaction deducted.

The advent of the Euro, will have a major impact on not just the nationals of countries that join. Multinational, European-based companies may change their currency requirements to the Euro, thus widening the impact of its introduction. But for most offshore banks, the Euro will represent a further currency in which to provide clients with a service. To some degree, these systems are already in place because the ECU is already an accepted currency for telegraphic transfer and cheques drawn on a multi-currency account.

The minimum investment in multicurrency accounts tends to be between £2,000 to £5,000. Interest rates offered on such accounts will vary and as a general rule will not pay interest if the account falls below the minimum deposit. There are exceptions to this; Cater Allen in the Isle of Man will pay reduced interest if the balance falls below the minimum.

PRIVATE BANKING

You could argue that if you are wealthy enough to need private banking services, then you are really getting your money's worth from offshore private banking. Offshore private banking services have the experience and knowledge to arrange a short list of schools in France for your children and book theatre tickets in New York and London, all while you are on a business trip in Hong Kong.

It is this high degree of customer care that is common to the best private banking services. Each private client is allocated what is usually described as a relationship manager, whose sole task is both to engineer and administer the smooth running of a client's account. He or she is a banking executive through whom you will conduct the bulk of your financial affairs. If required, relationship managers can and do travel all over the world to meet clients. Should the need arise, then the formation of trusts and companies will generally be part and parcel of offshore private banking services.

Some banks, however, concentrate solely on investment management. The type of investment management defined under private banking services will be first class discretionary management. You will be kept in constant contact with your portfolio and its performance. Your changing circumstances will be carefully monitored and reflected in any changes in your portfolio.

The ability to keep up with you when you move must be a significant factor when choosing a private bank. Citibank, which aggressively markets itself as a global bank, has 1200 branches in more than 40 countries. Merrill Lynch has offices in 45 countries, and is a member of 32 Securities Exchanges. Credit Suisse has 100 offices worldwide.

With today's advanced telecommunications, correspondence between bank and client can take place over the phone, fax, via e-mail or the Internet. Clients can fill out an indemnity form [insurance] allowing transactions to take place over the phone or fax. Passwords are used and security is watertight.

Useful technological gadgets allow clients on-line access to their accounts from anywhere in the world. You can view balances, make transfers, enquire about banking details such as interest rates, and order cheque books. The degree of the relationship often depends on the size of the account. Credit Suisse, for instance, requires a minimum of £500,000 (or currency equivalent) in spare cash.

However, there has been a trend in recent years to attract "emerging" high net worth individuals rather than simply the currently wealthy. This has seen a proliferation of banks offering some form of private banking service at a much lower entry level (£20,000) in the hopes that the account will grow as the client gets richer.

MORTGAGES

Buying property abroad is something most people dream of and it is both cost effective and more flexible to seek a mortgage from an offshore lender. There are two main uses of offshore mortgage facilities. First if you are moving overseas for a defined period of time then you may wish to keep your onshore property, rather than sell it. The ideal solution would be to let it. However, you may find that your current mortgage restricts you from doing so. Offshore lenders have a much more flexible attitude to such arrangements. If this were the most suitable option then moving your mortgage to an offshore lender would be the best option. Some lenders, such as Jersey-based Royal Bank of Scotland International even allow mortgages for investment purposes, allowing you to buy further property onshore to let out. If your onshore lender has an offshore presence then you may find it a fairly painless transaction to move your mortgage to the offshore lender. It is worthwhile pointing out that in doing so you may find you are charged a premium on the mortgage rate. Some lenders with both onshore and offshore mortgage lending facilities still charge different rates. Lloyds Bank is one that does not differentiate. Woolwich International is another lender that does not penalise offshore lenders. In addition, it offers the full range of current onshore mortgage products such as discounted and cash-back mortgages to offshore clients.

It is possible for expatriates to benefit from certain tax advantages by having an offshore mortgage. In the case of an offshore sterling mortgage, for example, any letting income earned from a UK property can be offset against all mortgage interest payments. Section 81 of the 1994 Finance Act also permits an offshore borrower to keep all financial accounts offshore.

The second use for offshore mortgage facilities is for a foreign holiday or retirement home. Most offshore lenders will offer mortgages in a range of currencies. For example, Banco Totto & Acores will offer mortgages for property in Portugal in sterling, US dollars, deutschmarks and Swiss francs. Abbey National, which has operations in both Italy and France, will arrange mortgages in the local currency. However, if you do plan to let out your retirement home then you will generally need to arrange this when you take out the loan or it will not be allowed. This restriction applies to both lenders named above.

It is usual for foreign currency loans to be granted on a lesser percentage of the property's value than is normally made available on onshore mortgages - for sterling mortgages think in terms of 60 - 80%. Foreign currency mortgage providers will often impose their own lending criteria that will alter according to market conditions. For instance, there may be restrictions on the amount a lender will advance depending on where the property is based.

Many lenders limit mortgages to shorter periods - 10 to 20 years - rather than the more usual 25 years available from onshore providers. Foreign mortgages are available in repayment or endowment form, with interest rates being variable or fixed.

Check whether it is a condition of the loan that you must buy buildings insurance through the lender. If this is not the case, it is worthwhile shopping around for the best deal.

Given the additional credit risk of a property being based in a foreign jurisdiction, offshore lenders may expect some form of collateral to protect the loans - usually in the form of insurance or pension policies or other UK-based assets.

SHARE DEALING

Unlike general portfolio management, which hinges on long term investment decisions, active share dealing requires more ongoing investment decisions. The use of an offshore centre for share dealing is therefore most advantageous if you are currently a long-term expatriate and have the benefit of a more flexible tax-planning regime.

Advances in technology have provided a major boost for offshore share dealing services. The scope for interactive service between broker and client is now available 24 hours a day. So if you are one side of the world and your broker is on another, the difference in time zones is becoming less of a problem.

Most offshore banks will offer clients a share dealing service, although there are a number of dedicated stockbrokers operating offshore who could be worth exploring.

While the main share dealing services available offshore tend to be no different from onshore services, you will generally find increased flexibility in, say, dealing in a wider range of currencies, derivatives and securities lending.

Execution only: This is the most basic service and is for investors who will make all the buy and sell decision themselves based on their own research. As the service simply acts on your instructions you really need to know what you are doing. Needless to say, this is the cheapest option in terms of cost.

Advisory Service: This is a step up from execution only, in that you still need to be fairly knowledgeable on markets but it gives you the ability to talk over your decisions with an expert. The final decision still rests with you.

Discretionary Services: This is the top end of the scale in terms of service and cost. You will be assigned a personal broker who will obtain a thorough knowledge of your investment profile, attitudes to risk and aspirations for the portfolio. Based on this information, the broker will then actively manage your portfolio. At regular intervals you will be contacted to ensure that everything is still on track in terms of what your portfolio is trying to achieve. This service is not generally available for portfolios under £150,000 ($250,000)

When you are potentially dealing from the other side of the world, the level of back office administration on your portfolio can be crucial. Many overseas investors arrange for share certificates to be held in what is called a nominee account. Essentially with these accounts, share holdings are pooled into one account held by the stockbroker's trustees of the Nominee Company. Nominee accounts came into their own when electronic settlement reduced the time permitted for making payments. Expatriates in far flung corners of the globe, and hindered by unreliable postal systems, have found such a service essential.

Offshore banks	Contact numbers
Altajir Bank, Cayman	(345) 949 5628
Abbey National (Gibraltar)	00 350 76090
Abbey National Treasury International (IOM/Jersey)	
	00 44 1624 662244/
	00 44 1534 885000
Alliance & Leicester International (IOM)	00 44 1624 663566
Allied Dunbar International (IOM)	00 44 1624 661551
Ansbacher (Jersey)	00 44 1534 504504
Bank of Scotland (Jersey)	00 44 1534 38855
Bank of Scotland (IoM)	00 44 1624 623074
Bank of Wales (Jersey)	00 44 1534 873364
Barclays, Channel Islands and IOM	00 44 1534 880550/
	00 44 1481 723176/
	00 44 1624 682000
Birmingham Midshires (Guernsey)	00 44 1481 700680
Bristol & West International (Guernsey)	00 44 1481 720609
	0800 (Toll free UK only) 833222
Britannia International (IOM)	00 44 1624 628512
C&G Channel Islands (Guernsey)	00 44 1481 715422
Cater Allen Bank (IoM)	00 44 1624 629201
Cater Allen Bank (Jersey)	00 44 1534 828000
Cheshire Guernsey	00 44 1481 726885
Cooperative Bank (Guernsey)	00 44 1481 710527
Derbyshire (IoM)	00 44 1624 663432
EBS Bank (IoM)	00 44 1624 670900
First National BS Guernsey	00 44 1481 710400
Flemings (Jersey)	0800 (Toll free UK only) 801767
Guinness Mahon Guernsey	00 44 1481 723506
Halifax International (IoM)	00 44 1624 612323
Halifax International (Jersey)	00 44 1534 59840
Hambros Bank (Jersey)	00 44 1534 815555
Hanson Bank	00 44 1481 723055
Hill Samuel Bank (Jersey)	00 44 1534 604604

Offshore banks	Contact numbers
Investec Bank (Jersey)	00 44 1534 283210
Irish Nationwide (IoM)	00 44 1624 673373
Irish Permanent (IoM)	00 44 1624 676726
Isle of Man Bank	00 44 1624 625533
Jske Bank	00 44 171 264 7700
Leopold Joseph (Guernsey)	00 44 1481 712771
Lloyds (Jersey)	00 44 1534 284000
Lombard Banking (Jersey)	00 44 1534 27511
Midland (Jersey)	00 44 1534 616000
Nationwide International (IOM)	00 44 1624 663494
National Westminster (Jersey)	00 44 1534 282828
Newcastle Bank (Gibraltar)	00 350 76168
Northern Offshore (IOM)	00 44 1624 629106
Northern Rock (Guernsey)	00 44 1481 714600
Norwich & Peterborough	00 350 45050
Portman Channel Islands (Guernsey)	00 44 1481 822747
Robert Fleming (IoM)	0800 (Toll free UK only) 289936
Royal Bank of Canada (IoM)	00 44 1624 629521
Royal Bank of Canada (Jersey)	00 44 1534 27441
Royal Bank of Scotland International (Jersey)	00 44 1534 24365
Schroders (CI)	00 44 1481 716961
Singer & Friedlander (IoM)	00 44 1624 623235
Skipton (Guernsey)	00 44 1481 727374
Standard Bank IoM	00 44 1624 662522
Standard Bank Jersey	00 44 1534 881188
Standard Chartered Bank (Jersey)	00 44 1534 507001
Sun Banking Corp (Jersey)	00 44 1534 608060
TSB Bank Channel Islands	00 44 1534 503909/ 00 44 1481 724061
Ulster Bank (IoM)	00 44 1624 672211
Woolwich Guernsey	00 44 1481 715735
Yorkshire Guernsey	00 44 1481 710150

Source: Moneyfacts

CHAPTER 8

INVESTMENT MANAGEMENT USING OFFSHORE PRODUCTS

Apart from the vast number of offshore funds you can access direct, there are a number of innovative products you can obtain offshore which use offshore funds as the underlying investment while offering an overall product which can cut down on investment administration, increase tax efficiency and are generally geared to reach a specific investment goal. The majority of these more innovative products are offered by life companies and range from relatively simple single premium and regular premium savings plans to portfolio bonds and guaranteed or protected products. Most of these products will include a life cover wrap. The performance of such products depends on the choice of the underlying investment. Many groups will offer a wide choice of in-house funds which are grouped according to risk profile. Increasingly, however, more and more groups are recognising that investors may not be happy to be limited to groups' in-house funds. This has seen a recent wave of life groups providing links to the funds of top performing investment groups.

Many insurance companies offering these products are based in the Isle of Man, the islands of Jersey and Guernsey, and Luxembourg. Isle of Man-based Royal Skandia was the pioneer in offering an extensive range of links to investment houses to be used as the investment base for its products. Other insurance groups which now offer outside links include Clerical Medical International, Sun Life International, Royal & Sun Alliance, and Scottish Provident International. A further innovation in terms of offshore life group products has been the increase in protected or secure products. These products will typically limit the downside of any investment. Scottish Life International, Sun Life International and Albany International offer protected products.

A brief description of a number of these products is outlined below. The choice of product will very much depend on your individual circumstances and your investment goal. The useful checklist at the end of this chapter will help sift through the product options when assessing your investment goals.

TYPES OF PRODUCTS AVAILABLE

Offshore Whole of Life

This is generally constructed as a flexible offshore unit-linked whole of life plan that aims to provide a cash sum on death. It will allow you to alter the balance of savings and protection when your needs change. It is usually possible to take out these plans on the lives of minors aged between 1 and 17 years. Such plans have a number of bolt-on options that are normally available at an extra cost. These include terminal illness benefit, term insurance benefit, accidental death benefit, waiver of contributions plus indexation. The investment element will comprise a choice of underlying funds offered.

Offshore Portfolio Bonds

An Offshore Portfolio Bond is constructed to hold most of your worldwide investments in one place making for ease of administration, cost savings and tax efficiency. It is basically a portfolio management service constructed as a single premium unit-linked life assurance policy. The specialist administration service provided ensures that you benefit from a worldwide portfolio of investments without the accompanying administration burden. In addition, it is possible to appoint a professional investment adviser to manage the bond. The adviser should have access to the latest information on markets and trading conditions, enabling a quick response to market changes and trends in order to maximise the earnings potential of the portfolio. Groups generally offer two versions of the offshore portfolio bond - managed and personal. A Managed Portfolio Bond enables policyholders to select from any UK-authorised unit trusts, US mutual funds and unit trusts authorised by other recognised territories.

For the more independently-minded investors, the Offshore Personal Portfolio Bonds provide an opportunity to hold a portfolio of stocks and shares, Government bonds, top performing mutual and unit trust funds, currency deposits and similar instruments that meet their own precise, individual needs. The scope for investment is very broad and by avoiding over-dependence upon one market, sector or country, investment risk is reduced. Investments that are currently in an investor's home country can be transferred into the Offshore Portfolio, which will increase the level of confidentiality.

For this administration service the investor is charged a set cost as opposed to having to contend with the uncertainty and unknown expense associated with undertaking the management of the portfolio personally. Dealing costs may also be reduced by virtue of an Offshore Life Insurance company's enormous purchasing power. These reductions in dealing costs are passed directly on to the investor.

It is important to weigh up any extra costs involved in having the life cover wrap on your portfolio as opposed to opting for a non-insurance linked portfolio management service. In addition, as far as UK investors are concerned, the UK has recently attacked the principle of a 'Personal Portfolio Bond'. In its March 1998 Budget, the UK unveiled a long-expected clampdown on personal portfolio bonds (PPBs) in the form of a penal taxation which attaches a deemed 15% per annual taxable gain on all PPB policies (existing and new) after 6 April 1999. The tax charge on this 15% deemed gain is in addition to the normal chargeable gains that apply to non-qualifying life policies. The deferred start date will effectively give those who have purchased these bonds "the opportunity to surrender them before the proposed annual tax charge arises," according to the UK Inland Revenue.

This change makes Personal Portfolio Bonds, which are highly personalised, an unsuitable investment option for UK residents.

Offshore Life Assurance

A choice of regular and single premium savings plans. Most life groups will offer a range of plans based on risk profile and flexibility. They can be tailored to meet retirement, investment, mortgage or education fees. The contracts will provide long term capital growth through access to a range of funds.

PRODUCT PROFILES

By way of explaining how the different products can be taken and adapted from group to group, we have included a series of product profiles highlighting the innovative products on offer. These profiles were correct as at 16 March 1998, but as product details do change from time to time, it is important to check on the details before you invest.

Scottish Widows International Jersey With Profits Bond: This is a sterling denominated bond which is a tax effective way of taking advantage of returns available from the stockmarket, with the added security of a with profits investment from a well known insurance group. The bond offers a bonus rate of 7% per annum plus a potential terminal bonus. The allocation varies between 101% and 102.5% for investors aged between three and 70 years old, depending on the size of the premium. The product is written as 20 individual policies. Surrender charges are 6% of the fund initially, 4.8% on the first anniversary and 3.6% on the second. Regular withdrawals are permitted, up to 7.5% per annum of the total investment. The minimum lump sum investment is US$17,500 or £10,000.

Scottish Life International's International Secure Investment Portfolio: This product allows investors to construct a balanced portfolio investment in funds of secure deposits and capital secure stockmarket-linked funds. It is available in both US Dollars and sterling. The portfolio consists of two types of investment funds. The Secure Savings Accounts is designed to meet the needs of investors who wish to improve on the low interest rates currently offered by international banks. Basically, it is a fund of offshore bank and building society deposits and is 100% capital secure. The Deposit Bonus Fund allows you to invest in a combination of deposit instruments and options. Investors can choose their personal level of capital security in a range from 95% to 100%. The quarterly bonus is paid when the FTSE 100 and S&P 500 do not fall over the period. The bonus payment is fixed in advance every quarter. The Stockmarket Bonus Fund offers investors exposure to the Nikkei index as well as the S&P 500 and FTSE 100 indices. The bonus payment is not fixed in advance; instead it is linked directly to the actual growth over the quarter on each of the world's three largest stockmarkets. A 'profit share' will be determined for each fund at the start of the quarter and will change depending on market conditions and the level of capital security chosen. The lower level of capital security, the higher the profit share. The minimum investment is £15,000 or $24,000 and there is no maximum. There is an 8% establishment fee taken monthly over the first five years. The annual management fee on the investment funds is 1.25% and 1% on the savings account. Scottish Life International has also launched a similar product based on individual stockmarket indices.

Scottish Provident International's Preferred Provider Portfolio: This is structured as a life assurance bond giving access to a range of funds offered by leading investment groups including Fidelity, Schroders and Perpetual. There are no dealing charges and it is available in all major currencies. Based in the Isle of Man, the product benefits from tax efficient growth and can be divided into 100 policies. An additional £65 is charged for administration. There is a choice of two charging structures. Structure A will charge a fee of 1.5% each quarter for the first 16 quarters and 0.25% thereafter. Structure B incurs a charge of 0.9% every quarter. The minimum investment is £30,000 and up to 10% of the portfolio can be withdrawn each year without charge.

Albany International's Safety First Fund: This is a protected asset fund accessible via Albany's single premium insurance products. The fund is a link to Scottish-based Edinburgh Fund Manager's Safety First unit trust. The fund invests in a portfolio selected from the 200 largest companies in the UK by market capitalisation. Between 95% and 97% of the assets are invested in equities, while between 3% and 5% is used to buy options that protect against falls in the fund price. The fund provides investors with a protected price, which is raised each time the fund price goes up 10% or earlier at the fund manager's discretion. If the market falls the price of a unit holding cannot fall below the prevailing quoted protected price. Any new protected price cannot be reduced for at least 12 months, whatever happens to the stockmarket. At the end of each protection period if the UK stockmarket has fallen the protected price may be reduced by a maximum of 5%. If the Safety First Fund is accessed via Albany International's Delta Account, there is a 1% establishment charge. Thereafter there is a £50 per annum administration charge, but Albany International takes no fund administration charges. Edinburgh Fund Managers take a 1.25% annual management fee.

Clerical Medical International's Premier Bond: This is linked to three offshore with profits funds denominated in sterling, Deutschmarks and US dollars. Each fund will invest in the equities and bonds of the country in whose currency it is denominated. These three funds offer guaranteed growth rates, which increases for those investments over £100,000. The bond also offers links to six managed and 37 other unit-linked funds, including trackers, international equities and risk-graded funds. Some 27 of these funds have the management outsourced to groups including HSBC, Credit Suisse, Beta Capital and Oppenheimer. The minimum investment in the bond is £15,000 with the minimum investment in any one fund being £1,500. The bond is segmented into 50 mini policies and there are loyalty units for those holding the bond for more than 10 years. Annual withdrawals of 10% of the value of the bond are allowed at no charge. The bond has an allocation rate of 100% and no bid/offer spread. There is an establishment charge of 0.12% per month of the bond's value for a period of five years. The annual management charge will vary between 0.5% and 2.75% depending on the funds chosen through the bond. If the bond is surrendered in the first five years, there will be a charge, varying from 9% in year on to zero in the sixth year. A market value adjuster may apply for withdrawals from the offshore with profits fund.

Your investment goals	Offshore life assurance	Offshore whole of life	Portfolio management	Offshore umbrella funds	Specialist offshore services
Retirement	Yes		Yes	Yes	
School Fees	Yes				
Mortgage Payments	Yes				
Returning to country of origin	Yes		Yes		Yes
Savings	Yes	Yes			Yes
Lump sum	Yes	Yes	Yes	Yes	Yes
Assets in a portfolio			Yes		Yes
Life cover	Yes	Yes			
Withdrawal option	Yes	Yes	Yes		
Protection		Yes			

Source: Old Mutual International

Investment goals

CHAPTER 9

OFFSHORE FUNDS

Once a decision has been made to invest in an offshore fund, the hard work has only just begun. Not surprisingly for an industry that spans the world, there is a plethora of different types of funds all of which claim to do different things.

You might reasonably ask how an offshore fund differs from an onshore fund? Quite simply, it is a fund registered in one of the internationally-recognised offshore centres outlined elsewhere in this book. By being based in an offshore centre, the funds can be managed within a tax neutral environment and will not be subject to most forms of income or capital gains tax. Dividends will be paid without deduction of tax, although investors should be aware that certain income received by the fund from the underlying investments might be subject to withholding taxes.

It should also be noted that investors from certain jurisdictions might not be eligible to invest in offshore funds. For example, US residents are not eligible to invest in an offshore fund unless the fund has Securities & Exchange Commission (SEC) approval. Some offshore fund providers such as Global Asset Management (GAM) have gone down this route, but most providers have preferred to steer clear of SEC requirements. Other jurisdictions take a slightly more flexible approach in that investors are allowed to invest in offshore funds. But for an offshore fund to distribute information about the fund, they have to be approved by the regulatory body. For example, in the UK, offshore funds recognised by the Financial Services Authority (FSA) are allowed to distribute information to investors.

A further difference is that offshore funds are subject to simplified regulation and generally designed for a more sophisticated investor. A less restrictive investment regime means they can often use borrowing to increase the performance of the fund and may also invest in instruments or markets that could be considered to have a higher 'risk' associated with them. This also means you get greater choice with offshore funds as they will invest in a wider range of asset classes which many onshore funds are restricted in doing. Such diverse asset classes include emerging market debt funds and exotic currency funds.

Major statistics providers, such as Standard & Poor's Micropal and Lipper Analytical, monitor more than 6,000 offshore funds split into nearly 200 different categories. While this may sound daunting, it must be remembered some of the categories only include a handful of funds. The Peruvian equity sector in fact only has one fund, the Peruvian Investment Company. At their simplest, the sectors are sub-divided into equity funds, bond and fixed interest funds, money market funds, and specialist funds such as warrants or commodities.

The number of funds is proliferating each year as investment markets and tradeable securities become more diverse, but it is worth knowing that the offshore fund concept is not new - the oldest offshore fund was launched in 1933 by Robeco, while Barings, now owned by the Dutch bank ING, launched two of its funds in 1957. The 1960s saw a spate of launches from well-known groups such as Barclays Bank, Mercury Asset Management (now part of the Merrill Lynch group) and Jardine Fleming. In 1969 Jardine Fleming launched its offshore Japan fund which was the forerunner for its Luxembourg range Fleming Flagship funds, and is now one of the largest individual ranges of offshore funds.

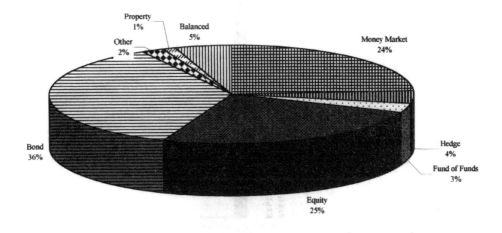

Property
1%
Balanced
5%
Other
2%
Money Market
24%
Bond
36%
Hedge
4%
Fund of Funds
3%
Equity
25%

Percentage of offshore fund assets globally by investment type

EXPLANATION OF THE DIFFERENT OFFSHORE FUND STRUCTURES

While there is a diverse number of funds to choose from, there are two basic offshore funds structures worth getting to grips with - open-ended and closed-ended. Both types of funds have equal attributes and investors must ask themselves which most suits their needs.

An **open-ended** fund does exactly what it sounds like to many investors. It is one where there is no limit to the number of shares or units the fund can issue at any one time. The fund manager or the investment company which runs the open-ended fund can decide to issue or cancel shares according to the normal laws of supply and demand which tend to dictate the price of individual equities. The advantage of the open-ended fund is that it allow investors to deal at any point in time irrespective of market conditions.

Closed-ended funds are the exact opposite of open-ended funds. As the name suggests, the number of shares or units is closed or limited according to a policy set by the investment management group which runs and markets the funds. Closed-ended funds tend to be more prevalent for investing in illiquid or esoteric markets or securities such as emerging market debt where it is relatively difficult for the fund manager to buy and sell the underlying securities. In this case, the fund manager does not want to be forced to deal in the underlying investments when investors want to move in and out of the fund because market conditions have gone against them. Some funds, particularly those which invest in illiquid markets, may impose a 'notice of redemption' period or only allow redemptions on one dealing day per month.

Additionally, closed-ended funds are likely to have a set lifespan from inception. This offers the investor some additional benefits. Firstly, the management group will often run the portfolio in a way designed to generate a set return at the end of the initial lifespan, allowing the investor to plan their personal investments around the likely return. However, they also know from the outset the length of time their capital will be tied up.

Such limited life funds tend to be less volatile in their early years, in terms of their share price movements. But, depending on the fund's performance, this trend can alter as the fund reaches maturity and speculators see the potential to make a quick 'turn' in the run-up to the end of the fund's life.

As a general rule of thumb, for an investor who would like to trade in and out of a fund to take advantage of market movements or who wants the security of knowing they can get out of a fund daily, an open-ended fund with its additional flexibility is more suitable.

The vast majority of these offshore funds are listed on recognised and established stock exchanges, making it easier for investors to buy and sell the funds, but also allowing them to track their performance and make comparisons with a relevant benchmark index. More than half the 6,000 offshore funds are listed on one or more of the 'big six' stock exchanges - Dublin, Bourse de Luxembourg, Cayman Islands, Bermuda, London and Hong Kong,

The advantages of buying a fund listed on one of these recognised exchanges - and some of the smaller exchanges in other areas of the world - is that there is clear transparent dealing and investors can have a higher degree of security over the fund they are buying into. It is a good gauge that a fund listed on a recognised stock exchange will have submitted to the listing requirements the regulatory bodies for the exchanges demand. This will ensure a greater degree of professionalism.

Over time, as the growth of offshore funds continues, more and more small exchanges in other Caribbean or Middle Eastern markets will be fighting to be the home for these funds. Investors should satisfy themselves the regulatory regime is rigorous enough to ensure that the listing requirements for funds do not admit any rogue vehicles.

While it is not too difficult to get to grips with the basic differences between closed-ended funds and open-ended funds, investors will find that there is little commonality between offshore fund structures generally. One of the reasons why different types of funds exist is that a fund promoter may decide that it wants to target a particular type of investor - say, European. It would therefore make sense to structure the fund in a format a European investor understands and prefers. In addition, each offshore centre has adopted its own legislation to enable fund launches. This has led to funds structures being called different names but achieving similar aims.

FCP (fonds communs de placement) is a collective investment vehicle based in Luxembourg. An FCP is often referred to as a Luxembourg unit trust and has the same flexibility and legal structures as a UK unit trust, making it popular with small, private European investors.

Limited partnerships - an investment company established as a partnership between a small number of investors, usually for tax reasons. They tend to be most common for hedge funds and specialist funds investing in higher risk areas such as derivatives and commodities where the partnership has a fixed life and is generally not designed to be open to a wide number of investors.

Open-ended investment funds are known as unit trusts in the UK and Hong Kong, mutual fund trusts in Canada and unit investment trust in the US. Unlike investment companies that have to abide by corporate law and submit reports to the Stock Exchange (see below), an investment fund only has to communicate with its investors and is not required to notify an exchange. Open-ended investment funds tend to be more common in the offshore fund arena than closed-ended funds. Amongst the open-ended fraternity, the two most common pieces of jargon which investors will come across are umbrella funds and UCITS (see below).

Open-ended investment companies are called mutual funds in the US; in Luxembourg they are called a SICAV (Societe d'Investissement à Capital Variable); in the UK and Jersey they are called an OEIC (Open-Ended Investment Company); and in Ireland they are known as VCIC (Variable Capital Investment Corporation). An investment company has to abide by the relevant company law regime, such as producing a report and accounts or reporting to the relevant stock exchange any information which may affect the performance of its shares. Any variations relate to the borrowing power of the funds, the constitution of the boards of directors and the role of trustees.

Closed-ended investment companies are called investment trusts in the UK, SICAF (Societe d'Investissement à Capital Fixed) in Luxembourg and FCIC (Fixed Capital Investment Company) in Ireland.

UCITS stands for Undertaking for Collective Investment in Transferable Securities and refers to the rules which allow for the authorisation, marketing, and selling of collective investment funds across borders of the member states of the European Union. In practice, if a fund has UCITS status, it can be sold in any member country in the Community as long as it abides by the local rules on marketing and promotion. It was designed so that a fund run from Germany could be bought and sold in any one of the members states, but in reality only a relatively small number of funds have been sold across borders in any meaningful way, and investors in local countries have tended to invest in domestically-run funds.

FUND PRICING

Along with a variety of fund structures, offshore funds may also have different classes of shares or units and have charging structures to accompany these different classes.

Generally, there are no more than two or three share classes per fund, which are usually A, B and I, where A and B shares are for the private investor and I is for the professional or institutional investor.

In the A share category, the minimum investment will vary starting at $1,000 or the equivalent for investment and aimed at the private investor with smaller amounts of money to invest. The B shares tend to be for the private investor with larger amounts of money, say $100,000 or above to invest. The charges on the two classes also tend to reflect the types of investor they seek to attract. The class A shares may have an initial sales charge or around 5% and an annual sales charge of say 1-1.5%, part of which is used to remunerate an investor's adviser. If the adviser does not take the full commission, it can be rebated back to the investor in the form of a reduced charge. For the class B shares there may be no initial sales charge and an annual charge of only 0.25%, which means the cost to the investor is considerably lower and any financial adviser may only receive a relatively small amount of annual commission for their ongoing advice. The share class for the professional or institutional investor will start with a minimum investment in the region of $1m. There will no be initial charge and an annual charge of only 0.1-0.2%.

The different share classes may also refer to a fund's ability to pay out or accumulate income earned from its investments. A fund that pays income is often referred to as a 'distributor' fund and one that accumulates the capital is called an accumulator or 'roll-up' fund. Sometimes, the A and B share classes may refer to the distribution and accumulation units.

In the UK, for a fund to qualify for distributor status it must distribute at least 85% of the available net income. A UK resident holder of a distributor fund will be charged income tax on the dividends received, but, on redemption will be charged capital tax on any realised gain and will be able to make use of any capital gains tax allowances or relief available.

A UK resident holder of an accumulator fund will be able to defer paying tax but will be charged income and gains tax once a gain is realised. Capital gains tax allowances or relief will not be available. More on this later.

Another key difference between offshore and domestically-managed investment funds is the basis on which the funds are priced - either on a dual or single priced basis. For example, in the UK unit trusts have two prices - the price at which units are bought - the bid price - and the price at which the fund manager sells units - the offer price. The difference between the two prices is called the spread, or more correctly the bid/offer spread, and this represents the real cost to an investor buying and selling units in a fund. With the increased globalisation of investment funds, the UK unit trust industry intends to move toward single priced funds by the year 2001, but until then it stands out as the exception on the pricing basis.

For most offshore funds there is a single price which is the same whether an investor wants to buy or sell shares or units, and it is much clearer and more transparent. Most offshore funds are priced on a forward basis. This means that the price is calculated after all the investments for that day have been received. In other words, the investor will not know the price at which the deal is struck until after the completion of the dealing day.

The advantage of a single priced fund, in addition to the simplicity in dealing, is that it is easier to find a quotation for such funds via either electronic media such as the Internet or the Reuters or Bloomberg terminals, or in international newspapers such as the *Financial Times*, the *International Herald Tribune*, the *Wall Street Journal*, or the *South China Morning Post*.

One of the most rapidly-growing area of offshore funds is hedge funds, which have attracted much notoriety and misunderstanding on the back of investors such as George Soros and his range of Quantum funds. Hedge funds have long since moved away from being the kind of vehicles which take wild punts on a rise or fall in the price of commodities or derivatives and are often glamorised in Hollywood style in films like Eddie Murphy's *Trading Places*. They are sophisticated trading funds that have no parameters and their only objective is to maximise returns. They are more fully explained later. (See page 119).

FUND SOLUTIONS

Investors are confronted with a plethora of fund acronyms and fund descriptions when assessing the range of offshore funds on offer. Conscious of the bewildering number of - and conscious of not wishing to lose investors to competitors - increasingly managers are endeavouring to help investors by offering solutions such as risk-rated fund ranges or a particular type of fund management approach designed to ease the burden of choice. Listed below are some of the more common types of fund ranges you will have to negotiate:

Fund of Funds

A fund of funds is a single investment fund that in theory invests in a range of top-performing funds either run by the management group or other fund management groups. The appeal and the sense of such funds is that an investor has access through one single fund to a range of funds spread geographically around the world. The investment group concentrates on picking the underlying funds where it believes the best investment returns are likely, thereby giving the investor an efficient, balanced portfolio using asset allocation methods to weight the investment. One of the disadvantages of this approach is that the investor is reliant on the fund or fund manager making the correct asset allocation predictions.

Multi-Manager Funds

The multi-manager approach offers a single route to a range of fund mangers from different investment groups. One of the pioneers of this approach has been GAM - Global Asset Management started by the mercurial Swiss Gilbert de Botton - which has used the style for over 15 years and uses more than 40 different managers.

The rule is simple. Where a group does not employ a fund manager with expertise in a particular area it contracts the fund management to a third party, but the investor goes into a fund carrying the name of the group they invested with. In most cases the investor will not pay a different level of charges if a group does have to use a third party fund manager and the agreement allows the investor to invest in a market which looks timely without waiting for the fund management group to launch a fund of their own. In theory, the investor should have access to top class fund managers across the board. This approach is reliant on the fund management group selecting the right third party fund manager. George Soros's Quantum Investment Group has been one of the most successful and longest running exponents of the multi-manager approach. For a management group it has the added benefit of being able to fire a poor third party fund manager at very short notice if they underperform. For an investor there is always the issue of whether it is prudent to be investing all your assets with a single management group, even if it is one which uses a third party fund manager for particular funds.

Umbrella Funds

Umbrella funds offer investors access to a wide range of funds offered, generally, by a single promoter. Basically, an umbrella fund is a collective investment scheme that has a number of sub-funds or share classes comprising a mixture of equity, bonds and cash in one large pool. Choices to switch between sub-funds or share classes are down to the investor and can generally be done at no or low cost. Depending on your tax status this may trigger a tax liability. The idea behind the umbrella fund is that it allows investors to take advantage of market changes quickly and efficiently. In the case of an umbrella fund offering only its own funds, there are certain drawbacks in that it does hinge on a group having sufficiently good range of underlying funds to be able to offer sufficient choice. In practice there are few groups anywhere in the world who have a sufficiently good wide range of funds to make this route a way of achieving top performance. They usually tend to specialise in particular areas of the world or types of assets. What an investor sacrifices in performance terms through this route they make up for with a low volatility and risk profile of the fund.

ROLL-UP FUNDS

A roll-up fund is a popular offshore fund choice because it does not distribute income or capital growth generally giving the investor the ability to defer paying tax. The following article from *Guinness Flight Asia* explains how roll-up funds work and gives some useful case studies. Although it is written with the UK investor in mind, the general explanation of roll-up funds is useful to other investors able to invest in this type of vehicle.

What is a roll-up fund?

Roll-up funds - often referred to as accumulation funds - derive their name from the fact that all income, as well as capital growth, is 'rolled-up' within the fund, and not paid out in the form of dividends. This accumulated income is, therefore, reflected in the price of a roll-up fund's shares. Accumulation funds were shaped by UK legislation introduced on 1 January 1984.

In the period immediately prior to the new legislation, investors in roll-up funds were able to convert income into a capital gain and, by using their annual Capital Gains Tax exemption, all tax liabilities could be avoided. As a result, the early 1980s saw a dramatic growth in the use of roll-up funds, particularly those investing in high income producing assets such as bonds and cash.

The legislation of 1984 instantly closed this loophole by introducing "distributing fund status" for those funds wishing to distribute income legitimately, and where the gross distributions of income were liable to income tax, for a UK taxpayer. For funds where less than 85% of the income received was distributed, then a new 'roll-up' status applied, such that all gains were charged to income tax. However, there remain considerable tax advantages available from the use of roll-up funds.

Roll-up funds are usually structured as open-ended collective investment vehicles, similar to unit trusts, although they can have a closed-ended structure (like Investment Trusts). They are generally domiciled in offshore centres such as the Channel Islands, Dublin and Luxembourg.

As well as certain tax advantages, roll-up funds offer a considerable administrative advantage to investors, who are able to avoid the proliferation of certificates associated with reinvested income. Additionally, the total return on an investment can be easily measured at any time, simply by comparing the current share price in the newspaper with the original purchase price.

Crucial to the tax advantages that a roll-up fund offers UK investors, is its location in a jurisdiction where the fund itself is not liable to tax on any income, or capital gains, arising from its assets. Location in a low tax jurisdiction does not mean that all income received by the fund is untaxed, as the fund may suffer some withholding taxes on investments contained in its portfolio. However, withholding tax only applies to equities, where income is a less significant part of the total return than with bonds or cash. With bonds and cash, income will be paid gross.

For a UK resident investing in a roll-up fund, there is no tax liability until encashment, in part or in whole. At this point, any gain realised by the investor will be liable to income tax at the marginal income tax rate of the investor. Because tax is only paid when a gain is realised, an investor is able to defer paying tax for long periods, thus enabling them to earn interest on higher sums of money than would otherwise have been the case. It also enables investors to put off paying tax until their personal tax rate falls; either because their income from other sources has declined, or due to statutory changes in tax rates.

Any losses arising when a roll-up fund is encashed constitute a 'nil income gain' and cannot be offset against other income tax liabilities.

Case 1: Regular withdrawal

If you require a regular income from your investments - paradoxical as it may seem - you may like to consider a roll-up fund, since withdrawals from a roll-up fund are more tax efficient than receiving interest payments. This is because each withdrawal is deemed to include both an element of repayment of the original capital and an element of accumulated income, with tax being levied only on the income.

If we assume that you have £100,000 to invest at a fixed interest rate of 5% and you pay tax at the higher rate, i.e. 40%, we can compare the difference between receiving £5,000 a year from a UK deposit account and £5,000 per annum by way of a withdrawal from a roll-up fund.

With the UK deposit account, all £5,000 is received in the form of income and, as such, is liable to tax at 40%. By contrast, the value of the roll-up fund investment after one year is £105,000, of which £100,000 is original capital and £5,000 is a gain potentially liable to income tax. Any withdrawal is treated as having been taken from both capital and income. With £100,000 of the £105,000 being principal, just over 95% of the withdrawal is deemed to be a repayment of principal, so only 5% of the £5,000 distribution is subject to tax. As an increasing amount of capital is withdrawn, the proportion of income in any withdrawal increases, as does the tax paid over time. However, even after an extended period, this remains significantly lower than that paid on a UK account.

The figures below show the amount of tax paid each year - using the same assumptions of £100,000 invested, 5% interest rate, 40% tax rate and £5,000 received gross of tax each year - from a UK deposit account and a roll-up fund.

For a UK deposit account, in each year, £2,000 of the £5,000 received is paid in tax, resulting in a cumulative tax bill after ten years of £20,000. With a roll-up fund, only £95 is paid in tax in the first year, rising to £772 in year ten, with a cumulative total tax payment of £4,556 for the whole period.

In this example, as the rate of interest received from the fund was fixed at 5%, the value of the underlying investment remains at £100,000 in both cases. Eventually the roll-up fund will consist entirely of income, as all initial capital will be deemed to have been encashed. At this stage the total tax borne by both funds has become the same.

Providing that your income requirement remains below the total growth of the fund, then the capital value of the holding will not have depreciated.

Case 2: Departing Overseas

Moving abroad also offers the possibility of reducing - and, in some instances, completely avoiding - paying tax on investment gains made on a roll-up fund. For a UK investor, this is after a period of five years outside the UK.

If you are intending to move abroad some time in the future - either to retire or because of your job - you should verify the tax treatment of roll-up funds in your intended destination. A number of countries still have tax regimes similar to the one which existed in the UK prior to 1984 where, as discussed earlier, profits made on roll-up fund holdings are regarded as capital gains, and are subject to taxation on a more favourable basis than income. To discover what the situation is, in the country to which you are moving, you should consult a tax expert there.

If roll-up funds are taxed favourably in the country in question, their inherent flexibility allows you to wait until you have taken up residency to crystallise gains. Consequently, in the right circumstance, your tax liability may be eliminated altogether.

Case 3: Temporary resident and non-domiciled in the UK

Temporary residence in the UK also offers opportunities for tax-efficient investment in a roll-up fund, providing any gains made on the investment are not remitted to the UK.

Anyone who is not domiciled in the UK - perhaps a foreign national who is resident in the UK, but does not intend to live there permanently - is only subject to taxes on income and capital gains from assets held outside the UK if the income or capital gain is brought into, or 'remitted' to, the UK.

For such a person, since roll-up funds are based outside the UK, even realised gains are outside the scope of UK taxation, providing they are not remitted to the UK. If income is not a requirement from any savings during the period of UK residence, temporary UK residents can use a roll-up fund to avoid subjecting investment income and capital gains to UK tax. It may also be possible to realise gains during the period of UK residence and avoid taxation in the country of domicile or next country of residence.

HEDGE FUNDS

Due to their aggressive investment stance and relatively complicated structure, hedge funds are the misunderstood investment creatures of the offshore world. However, they do offer some superb returns if used wisely. The following article by *Victor Hill, an adviser to Argyll Investment Management*, whose managing director, Richard Hills, has also published his own book on hedge funds called *Hedge Funds* (ISBN 0 9480 3528 5), which, is a good reference point for both novice and more experienced hedge fund investors.

‘

Introduction

The growth of hedge funds, with about $300bn of assets under management worldwide today, has been a silent revolution in the investment world. It has challenged fund managers and investment advisers to apply new and more sophisticated tools of analysis to produce better risk-adjusted returns for their clients.

The term "hedge fund" actually covers a multitude of sins, but all hedge funds can be said to be non-traditional investments. They are offshore pooled investment vehicles, or limited liability partnerships registered in the US, which invest in securities - rather than physical commodities or property - and which are permitted to use gearing, short-selling and derivatives to hedge market exposure and enhance returns. They engage in a number of strategies to generate consistently high returns over time, regardless of what the markets do.

Gearing is an idea which strikes fear into the hearts of more cautious investors, as it enables a fund to invest (and therefore possibly to lose) more than its entire capital. In practice, funds normally gear to benefit from two situations. First, for short periods where they identify exceptional investment opportunities but are unable or unwilling to unwind their existing positions to finance these new trades. Second, for longer periods where funds have found attractive, low risk arbitrage trades, but with relatively small returns. Here, by gearing up the overall position by, say, a factor of five, a much higher return can be achieved, even after any financing charges.

Short selling is an interesting concept and is critical to the understanding of hedge funds. How can you sell shares you do not actually own? Easily. You can borrow shares from someone who owns them, with the help of a stockbroker. The broker then sells them in the market in the usual way, depositing the cash sale proceeds into your account.

When you choose to instruct the broker to close out the position, the broker uses the funds in your account to purchase the shares you borrowed and replaces them in the account of the client from whom they were borrowed in the first place. In short, you agree a sale price at the beginning of the period and settle up at the end of the period at the prevailing market price. Thus, you make a profit if the stock price has fallen, and a loss if it has risen. At the end of the period in question - say, one, three or six months - your profit is equal to the start of period value of the shares you sold short, minus the end of period value, minus any cost of carry. So if the shares fell in price by more than enough to cover the cost of carry, then you are in profit.

This strategy is not entirely without its hazards. If at any time, while the contract is open, the broker runs out of shares to lend, then he or she can demand you deliver the required number of shares immediately - regardless of what the market price has done. These experiences of having to close out a position immediately whether ready or not are known as being "short-squeezed".

Say you have £100,000 to invest. You have a shrewd idea that 20 stocks are likely to go up nicely over the next three months so you go long (buy) these 20. But you are also fairly sure that 10 stocks are going down. If you were a long-only investor, you would shun them. But, following those clever hedge fund managers, you decide to sell these 10 short by £5,000 each. How much do you invest in the 20 that look promising?

If you said £5,000 per stock, you were wrong. Leaving aside settlement considerations, you have got £100,000 of capital plus £50,000 in cash you will receive by shorting the unattractive 10 by £5,000 each. So you have £150,000 to invest in the stocks that look a better bet (£7,500 each). Technically, the equation looks like this:

Net market exposure (%) = long position/short position/capital.

Let us say the 20 promising stocks rose by 10% over the three-month period and the weaker 10 fell by 10%. You made £15,000 on your long position (£150,000 x 0.1) and £5,000 on the short position (£50,000 x 0.1) making a total gain of £20,000 or 20% of your start-up capital. If you had remained a long-only investor, then your return would have been only 10%. So you have doubled your return. Clearly, gearing and short selling are intimately related. But whereas gearing in long-only funds entails the fund manager getting the market right (if the market falls they are losing borrowed money as well as their own), short-selling is more subtle, and in some cases, more forgiving.

The long-only fund manager who geared up to 50% in the example cited above, would, if the entire market had fallen by 10%, have lost 15% of their capital; but the losses on your long/short position would have amounted to exactly 10% - the same as for the market as a whole.

Thus funds which run long/short positions can be said to hedge market movements by taking positions which do not all move in the same direction. Moreover, the downside on adverse movements can be capped by means of equity options - a classic hedging device for those who run equity portfolios. True, some hedge fund managers actively speculate in the futures markets on the basis of their view of, for example, interest rate movements.

That does not, however, detract from the fact that hedge funds run long-short equity positions with or without caps, in order to enhance return and, at the same time limit downside risk. Many of the more mature hedge funds have been able to run equity portfolios over time, which have achieved consistent returns even when the Dow Jones was volatile.

The idea that funds can produce consistent absolute returns whatever the markets do is known as market neutrality. Funds which are market neutral cannot usefully be assessed by reference to a standard relative market benchmark or index such as the Dow Jones or the S&P Composite. They must instead be measured against a pre-determined target rate of return.

To add to the confusion, we now get to the exotic strategies. We have classified all of the hedge funds we track into six broad strategy categories, each of which can be refined by a number of investment styles and sub-strategies.

Long/short equity: This is the basic hedge fund strategy outlined above, with a predominant emphasis on equities. The objective of the strategy is market neutrality with enhanced returns. The "base case" long/short equity fund keeps net market exposure to 100% of capital at all times. More conservative long/ short funds mitigate market risk at all times by maintaining net market exposure at less than 100% of capital. Aggressive long/short funds may occasionally or even regularly amplify market risk by allowing net market exposure to rise above 100%.

Arbitrage: There are numerous types of arbitrage strategy, of which just two are set out here. Convertible arbitrage is when the fund goes long on a company's convertible securities and takes a short position on the underlying common stock. Fixed income arbitrage is where the fund exploits price differentials between fixed income securities and related derivatives.

Special situations: There are various types of strategy that focus on special situations.

- Distressed securities. Funds using this strategy tend to go long of the securities of companies on the verge of bankruptcy or undergoing restructuring. They may then use derivatives to neutralise market risk.

- Opportunistic fed income is the strategy of going long/short on debt issues that are perceived as mispriced on issue.

- Merger arbitrage is a strategy whereby a fund simultaneously purchases the stock of a company being acquired and sells short the stock of the acquiring company, thus making a directional bet that the deal will go through.

Emerging markets: Hedge funds active in emerging markets are really just long/short securities funds with a geographic focus in one or more near developed countries. Performance in this sector varies according to the region and country of specialisation.

Macroeconomic events: Macro funds seek to capitalise on regional and global economic events which impact on interest rates, securities, commodities and currency markets. They tend to be aggressive asset allocators who make substantial use of leverage and derivatives. The best known fund in this category is George Soros' Quantum Fund. The track record of this is well known, but strictly speaking, although considered by most analysts to be a hedge fund, Quantum and its peers are not really hedged funds. This is because they deliberately expose themselves to market risks in the expectation of getting their decisions right (at least more than 50% of the time). For this reason, we do not regard them as genuinely attempting to produce market-neutral strategies and hence we do not include such funds in our hedge fund portfolios.

Short-selling only: There are sometimes opportunities to profit from the sale of stock of overvalued, fundamentally flawed and fraudulent companies. Here, the fund focuses exclusively on the short sale of securities. Some funds, which take predominantly net short positions, hedge these positions with relatively small long holdings. This is a short based strategy. Most of the strategies outlined can be modified by the investment style of the manager. Some managers focus on medium-term stock growth, others are value investors; some keep trading to a minimum, while others are trading-oriented. Other sub-strategies include market timing, event arbitrage, opportunistic intervention and sector specific portfolio construction. In practice, most hedge funds can be said to be multi-strategic to some degree. Further, most hedge funds are passive investors, but some espouse shareholder activism and have even provided venture capital for new business start-ups.

OFFSHORE FUND PERFORMANCE

Once you have made an offshore fund selection, you will need to monitor its performance. Many offshore funds are now listed in international newspapers such as the *Financial Times*, *International Herald Tribune*, *Wall Street Journal* and *South China Morning Post*, where you will be able to keep up to date on fund price movements. For more comparative performance figures, you will need to consult specialist offshore magazines or statistics providers such as Lipper Analytical, Bloomberg, Standard & Poor's Micropal or Reuters HSW. The good news is that fund statistics providers - namely Lipper Analytical and Standard & Poor's Micropal - now have sites on the worldwide web where you can measure offshore fund performance, generally, for free.

There are a number of factors that affect fund performance as well as several ways of measuring it. There is no right or wrong way of measuring performance as each method has its advantages and disadvantages. The key point is to understand what is being measured and how.

Yield or Total Return? Yield is the calculation of the projected income of the fund as a percentage of the current price per share. The term "total return" is a measure of the per-share change in total value from the beginning to the end of a specified period, including distributions paid to shareholders. This measure includes income received from dividends, interest and capital gains. Total return is generally regarded as providing the best measure of overall fund performance.

Cumulative or Discreet? Fund performance is more commonly measured on a cumulative basis - which is the change in share price measured from the start of a specified period to the end of the specified period. This is normally over a one year, three year or five-year period. The advantage of this is it makes it easy to compare funds. The disadvantage is that it can hide a multitude of sins, particularly over longer time periods. For example, taking cumulative performance figures over a 5-year period could hide the fact that a top-performing fund achieved its spectacular performance in year five, negating poor performance in year 1,2, 3 and 4 to put it top of the tables.

This would suggest less than consistent fund management. Discreet performance looks at a fund's performance over each 12-month period. Fans of this method can then easily identify how often a fund appears in the top 25% of its sector. This can be a much more reliable method of consistent fund management.

Performance analysis. Examination of historical performance looking, initially, over discreet time periods is a much more meaningful measure than simply looking at, say, a three- or five-year track record. The problem with looking at cumulative performance is that it can mask periods of significant underperformance while, at the same time, showing strong results over the complete horizon. Cumulative data fails to highlight situations where a fund may have begun to underperform relative to an index or to the peer group. Discreet performance analysis can highlight more clearly and quickly any problems (and successes) that a fund is beginning to display. It is this discipline that can give early warning signals to find out why conditions are changing and whether the change is due to fundamental or technical reasons.

Peer or Benchmark? Offshore funds, like domestic funds, are put into appropriate sectors in order to measure their performance against funds that have similar investment aims. In addition, most funds will measure themselves against an appropriate index benchmark. For example, a US equity fund will generally benchmark itself against the S&P500 Index. It is important to look at both peer and benchmark comparisons, as measurement against one alone can be misleading. A fund beating its benchmark is of little comfort if every fund in the sector has achieved this and at a higher margin. A fund that achieves good performance in relation to its peers is cold comfort if the sector as a whole has underperformed the benchmark.

The main factor underlying good fund performance is good fund management and being in the right market at the right time. The latter is, unfortunately, more elusive than the first. The following article from **Michael Lipper and Christina**

Romas at Lipper Analytical looks at how funds performed during the last quarter of 1997. Also included are the top performing funds in a variety of sectors during 1997, plus the top 20 offshore funds over March 1996-1997.

6

The Asian Tiger Plays Dead

The fourth quarter of 1997 was difficult for offshore funds all around the world, according to the Lipper Overseas Fund Table (LOFT). As currencies in Asia were devalued in July, the momentum spread around the region and then across the globe. The most recent currency to take a hit fell the hardest: the South Korean won. In the past three months, the won lost almost half (-45.92%) of its value against the US dollar. This amounts to almost all of the won's depreciation in 1997; it was -50.06% vs. the US dollar for 1997. Other Asian currencies faced similar devaluation during the year, but did not take all their losses in the last three months: Thai baht (-46.61% for year, -25.53% for quarter) and Indonesian rupiah (-46.00% for year, -24.90% for quarter). The money market funds appeared to have relatively stable returns, -0.65% overall, since few focus on Asia. Within this grouping, the Indonesian money market funds in LOFT were -22.14% lower for the quarter.

Almost every geographic category was in negative territory for the quarter, equity funds in LOFT lost an average of -12.65%. This was punctuated by South Korea (-51.30%), and other Asian nations: Indonesia (-45.21%), China (-34.92%), Thailand (-34.09%), Malaysia (-33.04%) and Hong Kong (-31.02%). One fund that found profits in these down markets was Bermuda International's Global Manager - Hong Kong Bear Fund (+29.71%). As its name implies, this fund is designed to advance when the Hong Kong market declines.

Geo. Focus	Equity Funds	Number of Funds	Fixed Inc. Funds	Number of Funds	Money Market Funds	Number of Funds	Currency Change Vs US$
US	-2.64	301	1.59	190	0.48	141	N/a
Canada	-6.64	9	-3.08	24	-2.81	12	-3.44
Latin America	-12.32	82	-1.93	80	-	-	N/a
Brazil	-19.95	57	-2.05	30	-	-	N/a
Mexico	-6.47	4	-0.07	12	-	-	-3.77
Europe	-1.62	236	0.39	137	-0.10	24	-1.29
UK	1.86	95	4.19	69	3.69	49	2.00
Germany	-2.64	81	-0.60	204	-1.25	88	-2.14
Switzerland	3.45	55	0.30	58	-0.48	30	-0.94
Netherlands	-4.83	30	-0.67	38	-1.52	18	-2.22
Austria	-5.59	5	-0.46	3	-0.61	2	-1.60
France	-2.32	37	0.04	21	-0.83	30	-1.66
Belgium	-0.33	19	0.12	36	-0.80	30	-1.64
Italy	3.44	30	-0.03	23	-1.10	6	-2.74
Spain	-0.87	31	-0.24	22	-1.08	6	-2.43
Eastern Europe	-13.70	51	-2.77	8	-	-	N/A
Russia	-20.03	15	-20.99	1	-	-	N/a
Scandinavia	-8/86	14	-3.20	4	-	-	N/a
Sweden	-10.44	4	-4.36	12	-3.79	6	-4.67
Denmark	-	-	0.35	11	-0.65	1	-2.07
Japan	-19.54	175	-6.14	26	-7.55	16	-7.83
Pacific Region	-25.25	91	-5.16	6	-	-	N/a
Asia	-30.96	221	-11.03	11	-	-	N/a
Hong Kong	-31.02	43	-	-	1.14	5	-0.14
Malaysia/Singapore	-20.19	9	-	-	-	-	N/a
Malaysia	-33.04	12	-	-	-	-	-16.68
Indonesia	-45.21	21	-22.44	21	-22.14	1	-24.90
Philippines	-26.79	9	-6.5	1	1	1	-13.91
Thailand	-34.09	14	-	-	-	-	-25.53
China	-34.92	39	-	-	-	-	0.00
S. Korea	-51.30	3	-	-	-	-	-45.92
Taiwan	-20.69	24	-11.12	1	-	-	-12.38
India	-14.80	39	-4.40	1	-	-	-7.81
Australia	-16.62	31	-9.57	18	-9.58	10	-10.40
Dev. Markets	-17.73	97	-4.50	56	-	-	N/a
Global	-8.48	423	-0.06	371	-	-	N/a

4th quarter performance table

Italy and Switzerland are Exceptions

The main exceptions to this negative trend were funds investing in Italy and Switzerland. The Italian equity funds rose +3.44% during the quarter. December was the strongest month for Italy, showing gains of +7.27% compared to November's +1.13% up-tick and October's -4.84% decline. The most notable Italian equity fund is Sailor's - Italian Equity Fund (+12.73%). This fund is managed by the Italian firm, The Sailor's Advisory Co. At the end of December, most of the portfolio was in the Telecom industry (21.8%), Banking (21.7%) and Insurance (19.6%).

Swiss funds did well too, up +3.44% over the last three months. November's increase of +2.81% eliminated October's decline of -2.71%. This left all of the gain in December (+3.33%) for the quarter's positive performance. Credit Suisse Equity Fund - Swiss Blue Chip N (+9.63%) was the best fund within this sector. At June 1997 Pharmaceuticals & Chemicals accounted for almost half of the portfolio (42%). Insurance (14%) and Food & Beverages (10%) were the next largest sectors of investment.

Funds using Leverage Advance

Surpassing general equity funds were funds that used large amounts of leverage. Although down on average (-5.25%), leveraged portfolios still outperformed equity funds (-12.65%). Many of the best performing funds this quarter follow this risk-taking strategy. Magnum is the sponsor of the number one fund this quarter, Magnum New Markets (+40.00%) and the number nine fund, Trenton Small Cap A (+21.25%). The Magnum New Markets fund was fully invested in Azerbaijan at the end of the quarter. Trenton Small Cap A is a healthcare fund that had all of its assets in the US at the end of December.

% Total Return in US$	9/30/97 to 12/31/97
Magnum New Markets	40.00
Orbitex Dynamic Fund	31.62
Global Manager HK Bear	29.71
Freidberg Currency Fund	25.70
Europa American Option	25.48
Jaguar Fund NV-CI A	24.29
Seahawk International Inc	22.86
AHL Commodity Fund	21.61
Trenton Small Cap Fund	21.25
Gems Progressive – J2	21.01
Source: Lipper Analytical	

Top performing leverage funds

Other 'Leveraged/Hedge' funds in the quarter's top positions include the Jaguar Fund (+24.29) and Gems Progressive J2 (+21.01%). Managed by Tiger Management Corporation in New York, the Jaguar Fund's largest positions at the end of December were net long: US, French and UK equities and net short: Japanese and Hong Kong equities. The Gems Progressive Fund is a single manager fund, fully invested in the Jaguar Fund.

The outstanding performance of the Jaguar Fund is chiefly attributed to its large positions in US equities (75.8% long, 44.0% short). Although down slightly for the fourth quarter (-2.64%), funds investing in US equities turned in a respectable +20.56% return for the year. The US has become a more popular place to invest among offshore fund managers. Of the funds launched in 1997 that have been included in LOFT, 18% invest the US. This proportion of geographic breakdown is second only to 'Global' funds (25%). The proportion of new funds launched investing in the US has been growing steadily over the past several years.

For the quarter the US futures & options funds sector outperformed the US equity funds sector, up +1.31% versus down -2.64%. This is exemplified by number five fund, Europa America Option Fund (+25.48%) that outperformed its sister fund Europa American Equity (+6.49%). The two funds share the same portfolio, but the former uses options to enhance performance, while the latter does not.

US bond funds also produced positive returns in the quarter, up +1.59%. This is second only to UK bond funds, which gained +4.19%, boosted by the strength in the pound sterling (+2.00% vs. the US dollar).

Russia Best Over Year

Russian equity funds finished the year on top, despite a shaky end for 1997. Although declining on average -18.56% in the fourth quarter, the Russian equity funds almost doubled in value (+94.29%) over the year. (see chart below) Six of the overall top ten funds for the year focused on Russian equities. Finishing the year strongest among Russian funds is the Hermitage Fund (+228.17%). At year end, Hermitage Fund was most heavily invested in the Oil (44%), Electric utilities (16%), and Gas sectors (11%). Signet New Capital Markets C is up a very respectable +207.68% for the year. Following a few steps behind are Firebird Management's two funds: Firebird Fund LP (+141.18%) and Firebird New Russia Fund (+137.54%). The older of the two funds, Firebird LP was launched in May 1994 originally investing in the "voucher auctions". Since inception, Firebird LP has increased +415.16%. The year's number nine fund is the Clariden International Investment - Russia Equity Fund (+110.97%). At the end of December, 28% of its portfolio was invested in Lukoil, 21% in Unified Energy, 8.5% in Surgutneftegaz, and 5.8% in both Mosenergo and Rostelecom.

% Total Return in US$	12/31/96–12/31/97
Thornhill New Markets	403.20
The Hermitage Fund	228.17
Signet New Capital Markets	207.68
INVESCO Euro Warrant	157.76
Firebird Fund LP	141.18
Firebird New Russian Fund	137.54
Lancer Voyager Fund	118.69
Opportunity-Brazilian Value	116.10
Clariden International Investment – Russia	110.97
Russian Prosperity Fund CL A	106.16

Top performing funds in 1997

The Longer Term

Over the longer term, equity funds were still some of the best performing funds. Brazilian equity funds were strongest over five years, turning in +414.40%, far and away the best geographic region. Coming second and third over five years were Dutch equity funds (+193.31%), and Scandinavian funds (+191.30%).

The few exceptions were those areas that were also hardest hit this quarter: South Korea (-46.78%), Thailand (-46.04%) and Indonesia (-46.03%). Interestingly, their five-year performance was almost identical to their declines over the last three months. In South Korea, the last quarter's poor performance negated all of the funds' advances in the 1994-1996 period. (see chart below)

George Sores' funds continue to do well over the long term. Quota Fund was the number one fund over five years (+939.95%). Over the 10-year horizon, the Quantum Fund is the best performer, posting a +1419.97% increase.

Gold has reached its lowest point in 12 years, despite the frequent link between weak stock markets and a retreat into gold. This time it didn't happen. Gold-oriented funds have lost -30.23% during the final quarter of 1997, dragging the sector's 12-month average to -47.00%.

"

THE IMPACT OF CHARGES ON FUND PERFORMANCE

How much can charges affect an offshore fund's performance? The answer is significantly, but most of these charges are hidden beneath layers of small print which make it virtually impossible for the investor to calculate the true cost - the total expense ration (TER) - of a fund. The following article by *Richard Timberlake, co-founder of Fidelity UK; managing director of Portfolio Fund Managers and consultant on fund charges points*, out the effect of charges on offshore fund performance.

Introduction

Total Expense Ratios (TERs) are the only meaningful measure of the most significant part of a fund's charging system. The total expenses includes the disclosed management fees and all other expenses (particularly administration and custodian), billed on an annual basis against a fund's income.

Whilst these other charges are disclosed in a general legal sense, they are only available to investors after the event, within the income account and its notes, in historic cash terms, rather than shown easily, as percentage figures at the time of purchase.

Whilst some of the lessons have still not been learnt, many groups are now realising that setting annual "charges" high or low and just quoting them is meaningless. The TER, or total expense ratio, needs to be calculated, planned and managed. Groups with lower TERs are using this fact extensively in their competitive marketing and some groups, institutions and broker fund buyers are now seriously considering TERs before buying.

In Europe, we are still a long way behind the US in analysing TERs, developing logical competitive pricing policies and using this to commercial advantage.

TERs must be considered by sector and by size. Each fund or sub-fund TER should be looked at first in the context of all the comparable funds in that sector and secondly in the context of similar-sized funds.

Several groups still demonstrate inconsistency in their pricing management by being in both the lowest quartile for TER charges for some sub funds, but the highest for others.

Why "TERs" Are So Important

The stated investment management or investment advisory fee for unitised funds is sometimes 80% to 90% of the total annual fees and expenses billed to the fund and usually paid out of the fund's income, but in other cases it is only 40% or less of the total expenses. Thus current practice in Europe of disclosing the investment management fee as a percentage, but not disclosing the total level of fees in percentage terms (be they a little more or more than double), makes current disclosure practice and regulation meaningless; and the normal statement of management fees, especially for offshore funds, often downright misleading.

Small funds under $10m million and new funds are often prone to overpricing: - typically over 2.5% or 3% for equity funds. Investors should be wary of buying new offshore funds or specialist country funds, unless there is a stated TER limitation policy or "cap" (e.g. maximum charges or TER of say 2% or 2.25%).

The main expenses billed to an offshore fund, in addition to the management fee, are:

- Investment accounting, valuation and administration costs

- Custodian and sub-custodian charges (retainers and transaction costs)

- Shareholder records/registrar

- Company secretarial and organisation costs (sometimes included in fees)

- Directors fees and meetings

- Cost of report and accounts, interim statements, etc.

- Audit and Legal

- Domiciliary taxes and stamp duty

- Amortisation of Set Up/Reorganisation costs*

- Occasionally costs of marketing (literature, marketing support and even a proportion of telephone service centres).

One or two groups such as Regent, Jupiter and GAM also charge performance fees and, in these cases, the TER should be considered both before and after the performance fee. This is obviously an exceptional item, yet it is also clearly an expense payable by investors.

Note that it is common practice with offshore funds to charge the set up costs of a fund to the fund itself. Sometimes quite substantial sums (even over $1m) can be taken out of capital. These sums are then normally amortised over five years against the income account and form part of the TER. For a very large new "Umbrella Fund", with perhaps 5 - 15 sub-funds with different currencies, the cost of this structure can be a very significant expense billed annually as part of the TER. If the fund is promoted and expands rapidly, this expense usually becomes quickly under 0.2% per annum, but this is not always the case.

THE FEE MULTIPLIER

We call the amount by which the actual TER exceeds the stated management fee the "fee multiplier". Thus a fund with a 1% charge but a TER of 1.2% would be described as having a fee multiplier of 1.2x (20% over). The fee multiplier demonstrates the quantum by which the disclosed management fee is uplifted to get to the total expenses deducted from the annual income, or total expense ratio ("TER").

Because the expenses and charges to a fund over and above the investment management fee are mostly of a fixed rather than variable cost nature, the fee multiplier, and thus the TER, should decline as the funds get larger but this is not always the case and there are many exceptions where the economies of size are not being shared with investors.

In the case of offshore funds - in contrast to UK authorised unit trusts - few individual funds fit closely to the average and there is little general pattern - some funds having fee multipliers of 1.2x and others having multipliers of over 2.0x or 3.0x.

Because the annual investment management/advisory fee is widely stated and quoted in simple percentage terms, but the other expenses are only shown historically in the accounts (often sent to investors for the first time 15 months after they have invested) and these are then only shown in money terms, mostly in notes to the accounts, we believe that this is a major disclosure problem for the offshore funds industry.

In recent years, there has been considerable development of price competition and also changing approaches to charges in the European fund industry. Investors, encouraged by this competitive pricing, are now highlighting on the annual charge, particularly as it has become much higher in the case of onshore unit trusts and because several high charging offshore funds are now more actively marketed in the UK.

With the arrival of more sophisticated analysis, there is also a growing realisation that past performance is only one, possibly small, measure to be considered in fund selection, highlighting the handicap burden carried against future performance by those trusts and funds with significantly higher true annual charges (TERs) than others.

Whilst investors might reasonably assume that a fund charging 1% per annum is less expensive than one charging 1.5%, this is often not the case. For example, one well known fund "charging" 1% in fact has a TER of 2.8%, whilst another comparable fund apparently "charging " 1.5% in fact costs a true TER of only 1.8%. What appears half a percent cheaper is actually 1% dearer.

A simple example such as this demonstrates clearly that publication of management fees as an indicator of the "cost" of fund management is totally misleading. Management fees should not be published on their own, but total expense ratios should always be quoted.

LIMITATIONS AND SHORTCOMINGS OF TERS

A complication in any systematic analysis of offshore fund accounts at the moment is the absence of a regulatory framework and any standardised accounting and reporting procedures. This means that the reporting detail differs not only between offshore centres, but also within jurisdictions. This obviously makes detailed comparison difficult.

Any Expense Ratio Survey is nevertheless necessarily an imperfect product. Data is often not as current as it might be; given the lead-time required to publish audited annual reports. Funds may well have grown significantly since the accounting date. Current TERs may look different from those that are reproduced here, especially after strong market movements.

Whereas a management fee can be stated for the period ahead, a TER only relates to the last published accounting period, which may already on average be 15 months out of date. The ideal situation is where a management group, as in the US, will publicly state an expense limitation policy or a TER capping and state that the TER on its fund will not exceed a certain percentage. Such TER capping policies are to be encouraged.

Whilst TERs are much better indicators of cost than simply management fees, it is also true to say that certain items such as custodian handling charges and interest can be hidden in the capital accounts of the fund, whilst higher rather than lower stockbrokerage, which is invariably treated as a capital item, can be more important than many expenses billed to the income account. Similarly, slightly different accounting treatments in different jurisdictions or between different types of fund obviously reduce the validity of comparisons.

Total Expense Ratios (TERs) can only be considered in the context of other factors which may be of equal or greater importance in analysing a fund. The varying historic performance of equity funds, for example, will make small differences in TERs seem relatively unimportant. Nevertheless, when one looks forward there are often several equally promising funds offering the same prospect/hope of a return of say 10/15%, but the 3% per annum TER fund carries a very heavy handicap compared with the 1.5% TER fund.

However, we believe that there is already a sufficient degree of disclosure to merit a proper analysis of TERs. The industry's failure to analyse TERs in detail to date is not excused by the inconsistency in the detail of data and we should not be hiding from the vital conclusions thrown up by this exercise. There is no excuse for not doing TER comparisons. TERs still remain a far better indicator of cost than simply quoting management fees.

HOW TO PICK A WINNING OFFSHORE FUND

The disciplines needed to select a good performing offshore fund is, in practise, no different from the disciplines needed to select a good onshore fund. However, the diversity of the offshore fund industry does mean that investors need to be more vigilant and take a more sophisticated approach. The following article by *Paul Forsyth, managing director of offshore fund consultancy group, Forsyth Partners*, highlights the particular attributes needed to pick a winning offshore fund.

The analysis and selection of successful funds and investment managers involves a combination of disciplines. Some of the statistical databases that analyse individual funds and sectors are of the highest quality and are regarded as extremely useful. However, the most important factor in the fund selection process lies in the skilful interpretation and analysis of raw data. Essentially, this involves a qualitative appraisal. Firstly, let's take a look at offshore performance analysis and highlight some of the pitfalls or relying purely on numbers. To start with, which numbers do we look at?

Defining the Universe

This involves a comparative analysis of offshore funds with defined index benchmarks for a market or region or with the peer group of funds. The key point in finding out any comparative analysis is to compare like with like.

For example, in examining the universe of North American equity offshore funds, it is not valid to "lump together" those funds that focus on blue chip stocks, those that concentrate on mid cap and those that feature smaller companies. Equally, the universe may include warrant funds, special situation funds etc. Even terms like mid cap and small cap can be misleading as fund managers interpret them in different ways. Some funds identify stocks with market caps of US$500 million - US$1 billion as being small cap. Others set the range as US1 billion - US$4 billion. Obviously, the portfolios will differ. Finally, the databases frequently include "rogue funds" which are incorrectly included in a particular universe. The inclusion of these funds can lead to performance distortions and any universe needs to be carefully filtered.

Performance analysis

Examination of historical performance looking, initially, over discreet time periods is a much more meaningful measure than simply looking at, say, a three- or five-year track record. The problem with looking at cumulative performance is that it can mask periods of significant underperformance while at the same time showing strong results over the complete horizon. Cumulative data fails to highlight situations where a fund may have begun to underperform relative to an index or to the peer group. Discreet performance analysis can highlight more clearly and quickly any problems (and successes) that a fund is beginning to display. It is this discipline that can give early warning signals to find out why conditions are changing and whether the changes are due to fundamental or technical reasons.

New Funds

Just because a fund is new or has a limited performance record does not mean that it should be excluded from your 'buy' list as they can sometimes outperform established vehicles. This may be the case on occasions but the buying of funds simply because they are new is hardly a scientific approach. Situations where new funds can sensibly be considered could include country fund products ranges. For example, an investment house with a strong track record in Europe, perhaps demonstrated through a solid regional fund, may well have the capability of running individual country funds provided that its stockpicking skills are sound. If that company was known to be adding value primarily through asset allocation, it may not follow that its managers have the necessary skills to add value in country fund management. Similarly, a company that loses ground to the competition in regional fund management may be doing so because of poor asset allocation disciplines. Such a company may be more capable of managing individual country funds provided its stockpicking skills are good.

As well as looking at absolute performance, it is crucial to examine risk and volatility. We need to gain a clear understanding as to the risk that a fund manager is taking in achieving a specific level of performance. In its simplest form, volatility measures a fund's variance of each monthly return from the mean. The larger the figure, the higher the volatility of a fund and thus its risk. In examining volatility, the same disciplines used in looking at absolute performance are relevant. Again, the focus should be on discreet time periods for the reasons noted. Most statistics providers calculate and allocate a volatility rating to offshore funds, although this is not always made available to retail investors. However, increasingly specialist magazines are including volatility rating next to offshore fund performance figures. However, a typical example of the kind of funds and their associated risk ranging from low risk to high risk are: cash funds, fixed interest funds, balanced funds, equity funds and warrants funds.

Fund Size and Liquidity

Fund sizes can vary from a few million dollars (or even less) to sizes running into the billions of dollars. Both small funds and large funds have their advantages and disadvantages. Small funds to permit, albeit, by default, the manager to maintain a focused approach. On the other hand, such funds do not always give the facility to provide broad market exposure. For example, index tracker funds cannot be viable both in terms of achieving spread and in terms of cost unless they are fairly substantial in size. Giant funds do give investors comfort but in certain markets they can act as a constraint - and at times a significant one - on the investment manager. For example, a regional South East Asian fund sized at US$1 billion would be limited in its ability to shift, say, 5% of assets from Singapore into Thailand purely because of the size and liquidity of the underlying markets. This issue would arise even for a fund following a stock picking approach. The manager may be forced into holding a significant number of small weightings simply because he cannot buy his preferred stocks in sufficient quantity. In many cases, the giant funds end up "buying the market".

There are exceptions, of course. The substantial size of the US market and the liquidity that is a feature would generally not act as a constraint even to a fund sized in the billions of dollars. So, size is an issue but there are no simple rules to lead investors towards or away from large or small sized vehicles. Each fund needs to be looked at separately and an appropriate judgement made.

Understanding the Investment Process

The importance of the management group behind a fund or fund range, its resources and quality and the regulatory environment in which it operates are all vital factors to consider. Equally, the culture engendered by the senior management plays an important role in the successful management of client assets.

In looking at the investment aspects of an offshore fund management business, there are a number of factors to consider. These would include an analysis of the strength of the investment team, the quality of its individuals and their experience, the level of research undertaken by the company itself or the reliance which it places on third party information and whether monies are managed locally or at head office. The reliance on individual fund managers, as opposed to a team approach, is also considered with a view to assessing how vulnerable an organisation is to the departure of a specific fund manager. Identifying particular strengths which enables an assessment of whether an organisation is likely to be better than its competitors at managing specific types of funds. For example, strong asset allocation disciplines may be more important for a house that is heavily involved in the management of global or regional funds. Our research leads to the conclusion that the most successful asset allocation disciplines exist in those companies where the fixed interest team is strong. Their contribution, following their analysis of macroeconomic trends, can provide valuable input to the equity teams. Different skills are needed in the management of country funds, where stock picking is crucial. Investment houses can be categorised in terms of the general style they will follow in the management of portfolios. Typically, they may be described as top down or bottom up investors, value driven or a combination of a number of styles.

Fund Name	Latest Fund Size US$m	1 year	Rank	5 years
Insinger-Europ BD B	12.9	11.47	1	N/a
ABN Am Fd-Europa Bd	23.4	7.49	2	N/a
Obilex-28 MeditCur	13.3	0.46	3	N/a
Obiflex-L HiYd EurD	25.1	-1.11	4	50.48
DIT Eurozine	1966.5	-1.60	5	55.04
Obiflex-N ECU	10.9	-1.75	6	49.04
BBL Renta-High Yld C	75.4	-2.68	7	57.40
ML GCB-Eur BD USD B2	9.1	-2.78	8	N/a
Obiflex-34 MedCur D	1.4	-2.78	9	N/a
Eurorenta	2291.9	-2.90	10	56.45
DWS Europ Renten O	61.4	-2.95	11	N/a
MLBS Fixed Inc ECU	16.3	-3.07	12	44.70
JBaer MB-Europe BD B	487.7	-3.35	13	41.31
Grpe Indo-Eur Bond C	48.7	-3.46	14	46.91
Sogelux BD-Europe	83.6	-3.51	15	48.06
Mercury Sel-Eur Bond	11.5	-3.57	16	N/a
CS Bd(lux)ECU B	395.8	-3.60	17	47.07
Pleiade-Euro Bonds	22.7	-3.72	18	N/a
RBC Gibl-Euro Bd Inc	N/a	-3.74	19	N/a
DVG Fds-Europe Bond	0.4	-3.86	20	N/a
European Fixed Income Average	17231.8	-5.67	127	39.92

GLOBAL FIXED INCOME

Fund Name	Latest Fund Size US$m	1 year	Rank	5 years
Barclays IF-Glb Inc	14.0	19.95	1	92.77
Baring GUF-Hi Yld Bd	304.6	12.65	2	N/a
KML-II-High Yield	15.4	11.88	3	77.96
BT GAF-GlHi Yld Acc	93.4	11.70	4	N/a

Offshore fund performance figures

Deltec Widwd Inc Fund	33.6	11.13	5	74.84
LM Gib-Strateg Yld B	13.6	10.68	6	N/a
Mercury Bond	18.0	10.54	7	39.09
FrRussell-Hdg Gl Bd	79.2	10.37	8	N/a
Ashburton Fixed Inc	7.4	9.72	9	48.30
GT Strategic Bond B	-	8.93	10	N/a
GT Bond A	494.8	8.86	11	74.20
Deltec Spec Sits	26.4	8.83	12	87.94
ACMGL-Global Bond B	17.2	8.72	13	N/a
Parvest-GlBl 2 USD C	6.9	8.44	14	N/a
Ermitage Mgd-Income	0.7	8.34	15	N/a
Mercurcy Sele-US Gl Bd	81.6	8.28	16	41.80
Goldman Global Bond	61.8	8.25	17	N/a
ABN Am Fd-Global Bd	41.7	8.15	18	N/a
Lazard SYF-Strat Yld	6.5	8.06	19	N/a
MFS Amer-Char Inc B1	8.6	7.77	20	N/a
Global Fixed Income Average	**60625.6**	**-0.48**	**335**	**36.57**

EUROPEAN EQUITY

Fund Name	Latest Fund Size US$m	1 year	Rank	5 years
Dit-Aktien Europe AF	79.3	49.25	1	n./a
Skandifd Eq-Medit A	7.3	35.61	2	110.05
Euro Financial Eqty	32.0	34.52	3	N/a
Newton UGF-C Euro Eq	75.0	34.09	4	167.08
Carlson Eq-Medit	5.1	32.26	6	N/a
Scottish Value Euro Growth	1.7	32.23	6	N/a
BNP Inter Str-SthEur	57.6	31.61	7	138.38
Baring Europe Growth	544.1	30.08	8	158.18
UBS Eqinv-Sm Cap Eur	183.3	29.90	9	N/a
Fleming Front-Eur Discovery	93.2	29.80	10	N/a
Oyster Europe Value	17.8	29.26	11	N/a
ABN Am Fd-Europe Eq	174.3	29.15	12	N/a

Offshore fund performance figures

Hill Samuel Off European	19.3	28.80	13	139.85
Perpetual Off-European Growth	91.4	28.78	14	139.97
Albany Int. Eur STG	N/a	28.46	15	164.82
RG Europe	N/a	28.46	16	146.40
Bermuda Eq-European	24.40	28.08	17	115.15
UAP GMA-Europe Secs	81.4	27.62	18	132.91
GAM Pan European	204.5	27.22	19	N/a
Sun Life GL-Euro Growth	3.4	27.13	20	100.88
European Equity Average	**33184.1**	**18.13**	**210**	**126.07**

GLOBAL EQUITY

Fund Name	Latest Fund Size US$m	1 year	Rank	5 years
GP Global Technology	122.6	50.80	1	N/a
Henderson HF-Global Tech	46.0	40.80	2	N/a
Deltec Long Dis Eq	28.1	38.59	3	N/a
BBL(L) Inv-Health C	86.0	33.50	4	N/a
RWS-Akienfond-Verit	26.1	31.14	5	N/a
Conzett Europa-Invst	15.2	29.76	6	N/a
Adiverba	1119.1	29.36	7	98.35
DWS Telemedia	223.7	27.75	8	N/a
Energy Int	94.5	26.48	9	145.87
Malabar Intl	234.0	25.72	10	123.50
Credit Suisse Eq Fd Pharma	650.6	25.56	11	147.68
Titan GL Eqty Index	0.3	25.32	12	N/a
ACM Intl Health AX	18.7	23.07	13	106.82
Fidelity Gl Priv Tr	129.50	22.77	14	N/a
BBL(L) Inv-Telecom C	42.4	22.70	15	N/a
GAM (CH)Mondial	12.8	22.41	16	116.34
WPS Global Tech Fund	10.8	21.76	17	N/a
Buttress Equity Acc	13.8	21.65	18	46.34
Deutscher Vermogen I	8.1	21.35	19	100.45
Acm Intl Health B	50.1	21.26	20	N/a
Global Equity Average	**42562.9**	**3.49**	**373**	**71.99**

Offshore fund performance figures

ASIA/PACIFIC EQUITY

Fund Name	Latest Fund Size US$m	1 year	Rank	5 years
Discover Asia Invest	18.6	9.49	1	N/a
CIBC-CEF Pacif Ventr	8.2	5.82	2.	117.47
Capital Apprec GBP	5.2	2.84	3	44.77
Fleming FF Pacific	236.9	0.83	4	51.88
JF Pacific Securits	172.9	0.61	5	76.14
Capital Apprec USD	6.8	-4.41	6	36.73
ML Eq/Conv-Pacif Eq B	1088	-7.03	7	N/a
Fidelity Fds-Pacific	59.9	-13.51	8	N/a
Asian Capital Hldgs	345.0	-13.65	9	N/a
JF Pacific Income	44.6	-16.87	10	76.35
Mega Pacific	9.2	-18.22	11	18.06
Holland Pacific	89.0	-18.77	12	14.88
ABN Amro Far East	102.9	-19.12	13	30.70
Fidelity Far East	28.4	-19.54	14	-4.28
Fidelity Orient	16.0	-20.97	15	-5.84
Boston Pacific Grth	14.2	-22.71	16	5.53
Schroder Asia F East Growth	13.7	-22.78	17	38.81
Asian Technology B	16.9	-22.85	18	N/a
DIT Pazifik	421.9	-23.01	19	-11.64
ThorntonPacif Invst	3.0	-23.42	20	7.86
Asia/Pacific Equity Average	**3875.5**	**-28.02**	**93**	**18.08**

LEVERAGED/HEDGE FUNDS

Fund Name	Latest Fund Size US$m	1 Year	Rank	5 years
Thornhill Global Equity	4.2	403.20	1	N/a
Lancer Voyager Fund	18.3	118.69	2	N/a
TSG Leveraged USD	1.6	64.36	3	189.19
TSG Leveraged GDP	0.9	62.36	4	196.17
AJR International	5.5	61.26	5	N/a
Jaguar Fund A	8335.3	57.43	6	272.72
TSG Leveraraged CHF	2.9	55.20	7	175.27
Baron Capital Int	9.5	53.77	8	N/a
Quota Fund	1595.9	53.60	9	939.95

Prime Advisors	2.2	51.73	10	N/a
Alpha Atlas	25.8	51.49	11	N/a
Everest Cap Frontier	490.0	49.69	12	N/a
Cambrian Fund	20.0	48.56	13	N/a
Gabelli Int. A	18.9	48.43	14	185.84
Regan Int Fund	50.8	46.39	15	N/a
Odey European USD	25.3	45.84	16	N/a
Regent Pacific Hedge	65.0	41.44	17	N/a
Aventine Int	48.5	39.24	18	N/a
Explorer Fund	668.0	36.91	19	N/a
ACM US Gr Strat I	76.0	36.83	20	N/a
Leveraged/Hedge Fund Average	**37665.6**	**12.93**	**216**	**98.34**

Performance is given in total return figures for one year (12/31/96 to 12/31/97) and five years (12/31/92 to 12/31/97). Source Lipper Analytical

Offshore fund performance figures

CHAPTER 10

OFFSHORE CENTRE INFORMATION

INTRODUCTION

Below are comprehensive details on a selection of offshore centres. The information is designed to give a good overview of what each centre offers in order to help investors compare and contrast. In undertaking the task of verifying the information, it was noted how helpful each centre was in sending back information on time, the depth of response and extra information supplied. Those centres to be commended for courtesy and helpfulness include: the Isle of Man, the British Virgin Islands, Cayman Islands, Cook Islands, Gibraltar, Turks & Caicos, Switzerland, Liechtenstein and Guernsey. While both Dublin and Jersey replied, the information was standard rather than specific, which is of little help to investors. Information on all centres was supplied by Bloomberg/International Investment.

We have not covered every centre in this section. Contact addresses for other centres are included at the end of the section.

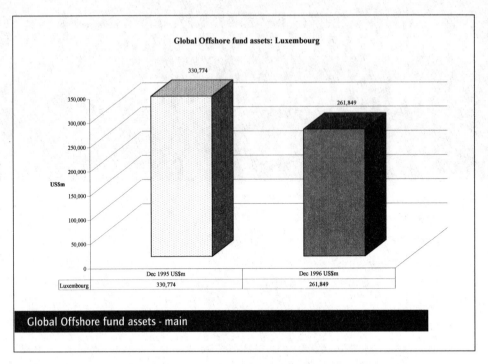

Global Offshore fund assets - main

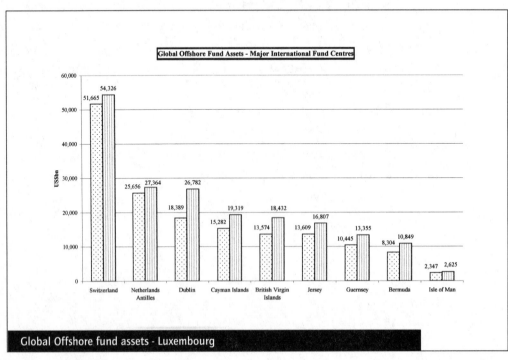

Global Offshore fund assets - Luxembourg

BAHAMAS

Time Zone: GMT - 6 hours (5 hrs April to October)

General

Political stability: Excellent	Bank secrecy law: Yes
English common law: Yes	Company type: IBC
Exchange controls: Nil	Offshore revenue tax: Nil
Tax treaties: Nil	Owner disclosure: No
Language: English	Domicile migration: Yes
Offshore Banking: Yes	'Shelf' companies: Yes

Corporate Requirements

Local Requirements

Minimum shareholders: 1	Registered office/agent: Yes
Minimum directors: 1	Local directors: No
Corporate directors: Yes	Directors disclosed: No
Secretary required: No	Local Meetings: No
Bearer shares: Yes	

Trusts

Availability: Yes

Asset Protection Trusts: Yes

Perpetuity: Life + 21 years

BAHAMAS GENERAL INFORMATION

Taxes

There are no taxes on income, earnings, capital gains, profits, estates, inheritance, or probate. Also, there are no withholding taxes on dividends, interest, royalties, or payroll

IBC Features

The International Business Companies Act of 1989 is very similar to the International Business Companies Act of the British Virgin Islands. Companies can be incorporated under this Act as long as they do not carry on a business in the Bahamas. Other than a local office, they cannot own real estate in the Bahamas and may not be a bank, an insurance or re-insurance company or a Trust company. A Bahamian IBC must maintain a registered office address in the Bahamas and there must also be a registered agent who must be a lawyer, solicitor or law firm, a licensed trust or management company.

There must be at least one director who may be a natural person or a corporation. Alternate directors can be appointed and other officers such as company secretary can also be appointed. The names and addresses of the directors and officers are not available to the public although this information may be lodged with the Registry if an official Certificate is required. The company must however maintain a register of directors at the registered office. The minimum number of shareholders is one and may be individuals or corporations. All issued shares must be paid for in full. The company must maintain a register of members at the registered office, although this information is not available to the public. The company is not required to file accounts or file any annual returns. The directors may hold meetings in the Bahamas or in any other country and there is no requirement for an annual general meeting. Subject to name approval, a company can be incorporated within 24 hours. 'Shelf' companies are available for immediate use.

US Treaty

* Limited Exchange of Information Agreement signed with the US Department of Justice under Mutual Legal Assistance Agreement which calls for exchange of information on tax fraud and laundering of funds but not offences which are purely tax-related.

Location

The islands of the Bahamas are located 50 miles off the US Florida coast and northeast of Cuba. It is in the same time zone as Miami, New York, Montreal, and Toronto.

Government contacts: The Registrar General's Department, Registry of Companies, P.O. Box N-665 Nassau, Bahamas, Tel: 809-322-2145. Fax: 809-322-5553

BARBADOS

Time Zone: GMT -4 hours (-5 hours from April to October)

General

Political stability: Very good	Bank secrecy law: Yes
English common law: Yes	Company type: IBC
Exchange controls: No (for IBCs)	Offshore revenue tax: 1-2.5%
Tax treaties: Several	Owner disclosure: No
Offshore Banking: Yes	Domicile migration: Yes
'Shelf' companies: Yes	Language: English

Corporate Requirements

Minimum shareholders: 1
Minimum directors: 1
Corporate directors: No
Secretary required: Yes
Bearer shares: No

Local Requirements

Registered office/agent: Yes
Local directors: No
Directors disclosed: No
Local Meetings: No

Trusts

Availability: Yes
Asset Protection Trusts: No
Perpetuity: Life + 21 years

BARBADOS GENERAL INFORMATION

Taxes

Although Barbados does have domestic taxes, the offshore financial sector pays a maximum rate of 2.5 percent. There is no capital gains or inheritance tax. Trusts are taxed at personal tax rates and no personal allowances can be claimed.

Confidentiality

Under the Offshore-Banking Act, client information of an offshore bank is kept confidential. Although the names and addresses of beneficial owners of companies must be revealed to the Central Bank, it will keep them confidential. However, it is not a criminal violation for government officials or other employees to reveal such information.

The Companies Act does not permit bearer shares, but shareholder anonymity can be attained by using nominees to hold shares, using a foreign trust to hold shares in a jurisdiction with secrecy laws, or using share warrants.

Barbados Tax Treaties

Canada	Norway	Switzerland*	United States
Finland	Sweden	United Kingdom	

* Switzerland's 1954 treaty with the United Kingdom is, with some differences, still applicable to Barbados and other former and present British territories. As a rule, the treaty does not cover dividend and/or interest payments from or to these territories.

Location

The easternmost island in the Lesser Antilles of the Caribbean, Barbados is located several hundred miles northeast of Venezuela.

Contact: Ministry of Finance and Economic Affairs, Government of Barbados, Treasury Building, 6th Floor, Bridgetown, Barbados, WI. Tel: 809-426-3815

BERMUDA
Time Zone: GMT -4 hrs (-5 hrs April to October)

General

Political stability: Excellent	Bank secrecy law: No
English common law: Yes	Company type: Exempt
Exchange controls: Nil	Offshore revenue tax: Nil
Tax treaties: Nil	Owner disclosure: Yes
Language: English	Domicile migration: Yes
Offshore Banking: No	'Shelf' companies: No

Corporate Requirements

Minimum shareholders: 3	Registered office/agent: Yes
Minimum directors: 1	Local directors: Yes
Corporate directors: No	Directors disclosed: Yes
Secretary required: No	Local Meetings: No
Bearer shares: No	

Local Requirements

Trusts
Availability: Yes
Asset Protection Trusts: Yes
Perpetuity: 100 years

BERMUDA GENERAL INFORMATION

Taxes
No withholding, capital gains, transfer, wealth, gift or income taxes for non-resident entities.

Exempt Company Features

An exempt company is generally one incorporated in Bermuda which has been exempted from the requirement of 60% Bermudan ownership, and may conduct business abroad from within, but not domestically within, Bermuda. An exempted company is free from exchange control restrictions and can therefore hold its assets in, and trades in, any currency except Bermudan dollars. In addition, there is no form of corporate income tax or tax on profits.

The Companies Amendment Act 1996 permits an exempt company to have as a minimum either two resident directors, or one resident secretary and one resident director, or one resident secretary and one resident representative.

There are more than 7,000 exempted businesses registered in Bermuda, with approximately 250 of them occupying offices and employing staff. The rest do not have a physical presence there but are managed by professionals in Bermuda.

Incorporation fees are generally higher than in other jurisdictions and there is a screening process of applicants. For this reason, Bermuda is a prestigious offshore business centre that attracts a very polished clientele and is popularly regarded as the world's 'cleanest' offshore financial centre.

Bank Secrecy

There are no bank secrecy laws, but the government will not allow disclosure of confidential bank information in cases strictly involving tax evasion.

** Bermuda has a limited Exchange of Information with the US that provides limited exchange of information on tax fraud and laundering of funds but not for offences purely tax-related.

Stock Exchange

Bermuda has its own stock exchange.

Location

Bermuda is located in the Atlantic Ocean approximately 680 miles east of the mid-Atlantic coast of the United States. Its time zone is one hour earlier than that of New York.

Government contacts: Ministry of Finance, Government Administration Building, 30 Parliament Street, Hamilton HM 12, Bermuda

Other Contacts: Bermuda International Business Association (BIBA), 41 Victoria Street, Hamilton HM12, Bermuda. Tel: 1-441 292 0632

BRITISH VIRGIN ISLANDS

Time Zone: GMT -5 hrs (no daylight saving)

General

Political stability: Excellent	Bank Secrecy: No statutory legislation. Governed by English Common Law.
English Common Law: Yes	Company Type: IBC
Exchange Controls: Nil	Offshore Revenue Tax: Nil
Tax Treaties: 2	Owner Disclosure: No
Language: English	Domicile Migration: Yes
Offshore Banking: Yes	'Shelf' Companies: Yes
Mutual Funds: Yes	

Corporate Requirements

Minimum shareholders: 1

Minimum directors: 1

Local Requirements

Registered office/agent: Yes

Local directors: No

Corporate directors: Yes   Directors disclosed: No

Secretary required: No   Local Meetings: No

Bearer shares: Yes

Trusts

Availability: Yes

Asset Protection Trusts: No

Perpetuity: 100 years + "wait-and-see" rule

BRITISH VIRGIN ISLANDS GENERAL INFORMATION

Taxes

Zero tax jurisdiction for offshore entities

Tax Treaties

Switzerland's 1954 treaty with the United Kingdom is, with some differences, still applicable to the British Virgin Islands (BVI) and other former and present British territories. BVI also has a treaty with Japan through the United Kingdom. Despite the existence of these two treaties, since the introduction of the IBC legislation, the BVI is rarely used for treaty shopping purposes. Accordingly, persons who have a genuine interest in these are strongly advised to contact tax experts in their own jurisdiction.

* Limited Exchange of Information Agreement signed with the US Department of Justice under Mutual Legal Assistance Agreement which calls for exchange of information on tax fraud and laundering of funds, but not offences which are purely tax-related.

IBC Features

Since the original legislation was introduced in 1984, more than 250,000 International Business Companies have been incorporated, and recently incorporation of new companies was at an average level in excess of 3,000 per month. The popularity of IBCs is a result of the flexibility of the IBC legislation combined with the advantageous features of the BVI as an offshore jurisdiction.

The features and requirements of a British Virgin Islands IBC include:

- It is entirely free of all BVI taxes and only pays an initial registration fee and thereafter a fixed annual license fee.

- Directors, shareholders, administration and the assets of the company may be within or outside the BVI.

- Formation requirements and ongoing administrative requirements are simple and efficient.

- Optional registration for public record of directors, shareholders and mortgages and charges, may be made if the company desires.

- Flexibility is maximised by provisions in respect of company name, the availability of bearer shares, the provision of nominee director services by licensees, the ability to hold meetings when necessary and the statutory recognition of electronic media as an acceptable means of conducting corporate meetings.

- Financial records may be maintained in any jurisdiction, and an audit is not required.

- There must be a registered office in BVI and there must also be a registered agent who has to be licensed by the Financial Services Inspectorate to provide trust or company management services.

- There must be at least one director who is appointed initially by the subscribers and the director(s) can be either natural persons or corporate entities. The names and addresses of the directors are not available in any public register although a company may choose to lodge such information with the Registry so that official certification can be provided if required.

- The company must maintain a share register at the registered office but it is not mandatory to file a copy with the public registry.

- Subject to approval, an IBC can be incorporated within 48 hours. Ready-made 'Shelf' companies are available for immediate use.

Trusts

Trusts need not be publicly registered and there are no requirements to file financial statements or other information.

Mutual Funds

On January 2, 1998 the British Virgins Islands brought into force the Mutual Funds Act 1996. The Act, among other things, requires all mutual funds to be registered or recognised and their managers and administrators to be licensed. Under the Act there are three types of mutual funds - private funds, professional funds and public funds.

Additional Services

Several Licensees are able to offer additional corporate services such as insurance management, trust formation, mutual fund management, and registration of Limited Partnerships and registrations of Ships.

Investor Protection

At present there is no investor compensation scheme in the British Virgins Islands. However, the Financial Services Inspectorate is considering legislation regulating financial service intermediaries at which time it may be possible to include provisions which may provide recourse to investors.

Where an investor has a complaint against a professional service provider in the British Virgins Islands, that complaint should immediately be drawn to the attention of the Director of Financial Services of the Inspector of Banks, Trust Companies and Company Managers. Only professional bodies of high standards and competence are licensed by the Financial Services Inspectorate and it is a condition of every license that its holder will perform in a manner and to a standard which will promote the BVI as a sound and reputable jurisdiction in which to do business. The matter becomes more complex in the case where an investor has a complaint against a BVI IBC. An IBC company (as opposed to a professional service provider) is not able to do business in the British Virgins Islands. For added protection, clients should ensure that the IBC is subject to local regulation in the place where it is carrying out its business.

Money Laundering

In keeping with the International trend of extending anti-money laundering legislation to include 'all crimes' activity, the British Virgins Islands passed the Proceeds of Criminal Conduct Act in 1997, which came into force on January 2 1998. The Reporting Authority under this Act is presently being constituted and guidelines are in the process of being formulated. These guidelines will give, among other things, clear instructions as to the filing of suspicious transaction reports by financial institutions in the BVI. To date, no prosecutions have been brought under the Act.

Location

The British Virgin Islands are located 60 miles east of Puerto Rico in the Caribbean Sea.

Government contact: Mr Robert Mathavious, Director of Financial Services, BVI Financial Services Department, Government of the British Virgin Islands, Road Town, Tortola, BVI Tel: 809-494-4190, Fax: 809-494-5016, Web site: **www.bvi.org**

Other contacts: Mr Urs Stirniman, President BVI Bank Association, c/o VP Bank (BVI) Ltd, PO Box 3463, Road Town, Tortola, BVI.

Ms. Rosa Rostrep, President, Association of Registered Agents, c/o Arias Frabrega & Frabrega Trust Co BVI Ltd., PO Box 985, Road Town, Tortola, BVI.

CAYMAN ISLANDS

Time Zone: GMT -5 hrs (-6 hrs April to October)

General

Political stability: Excellent

English Common Law: Yes

Exchange controls: Nil

Tax treaties: Nil

Language: English

Offshore Banking: Yes

Bank secrecy law: Yes

Company type: Exempt

Offshore revenue tax: Nil

Owner disclosure: No*

Domicile migration: Yes

'Shelf' companies: Yes

Corporate Requirements*

Minimum shareholders: 1

Minimum directors: 1

Corporate directors: Yes

Secretary required: No

Bearer shares: Yes

Local Requirements*

Registered office/agent: Yes

Local directors: No

Directors disclosed: Yes

Local Meetings: Yes

*Different criteria would apply to banks, trust companies, insurance companies, mutual funds, fund administrators and company managers.

Trusts

Availability: Yes

Asset Protection Trusts: Yes

Perpetuity: 150 years + "wait-and-see" rule

CAYMAN ISLANDS GENERAL INFORMATION

Taxation
Zero tax rate.

General
Specific advantages to the overseas investor, banker or company include the following:

- The stability of a well and progressively-governed colony of Britain.

- Excellent telecommunications, employing the latest technology, efficient air services, provided by five international airlines, reliable shipping links providing regular freight supply.

- Proximity to the United States - about an hour to Miami and good daily connections worldwide.

- A highly-developed infrastructure, including ports, airfields, roads, utilities and medical and educational services.

- A pool of lawyers, accountants and other professional highly qualified and experienced in the administration of financial services, applying their skills through institutions of repute observing government-set standards of probity.

- An advanced land registry system, with all titles guaranteed by Government, allowing quick and efficient transactions in property.

- No restrictions on the foreign ownership of real estate.

- Efficient processing and registry systems for banks, companies, mutual funds, fund administrators trust and insurance companies, making increasing use of Government's large computer system.

- While allowing for some flexibility, banks, trust companies, insurance companies, mutual funds and mutual fund administrators are monitored by qualified and professional staff in the various divisions of the Cayman Islands Monetary Authority to see that they adhere to standards equal or higher (e.g. capital requirements) than internationally accepted standards.

- Ready access to an established financial centre with lines of communication with Eurocurrency markets.

- Absence of direct taxation, providing the ability to accumulate profits free of tax and, for branches or subsidiaries of parent companies elsewhere, the ability to repatriate profits at most advantageous times.

- Absence of exchange control, allowing the movement of capital in and out of the colony and the opening and maintenance of accounts in any leading currency.

- Severe penalties ensure confidentiality between client and professionals under the Confidential Relationships (preservation) Law, qualified only by Government's accepted duty to assist in the investigation of drug-related and other serious crime.

- Well-staffed government departments, backed by computerisation, allowing ready availability of information and relevant legislation.

- Sophisticated private sector, providing such facilities as courier services database access, and facsimile transmission.

- Fixed rate of exchange: CI$1 = US$1.20. Exchange rate in relation to other world currencies floats with US$.

- Modern commercial legislation and legal and judicial systems based on that in the United Kingdom

US Treaty

*Limited Exchange of Information Agreement signed with the US Department of Justice under Mutual Legal Assistance Agreement which calls for exchange of information on tax fraud and laundering of funds but not offences which are purely tax-related.

Stock Exchange

The Cayman Islands has a stock exchange.

Investor Protection.

Other than the Preferred Payments scheme for depositors in Category "A" banks, there is no investor's compensation scheme.

Location

The Cayman Islands are located in the British West Indies due south of Cuba, 475 miles south of Miami, Florida, and 200 miles west of Jamaica.

Government contacts: The Monetary Authority, PO Box 10052, Elizabethan Square, Grand Cayman. Tel: (345) (49 7089. Fax: (345) 949 2532.
Cayman Islands Government Information Services, George Town, Grand Cayman. Tel: (345) 949 8092. Fax: (345) 949 5936.
Web Site: **www.cimoney.com.ky**

Other contacts: Cayman Islands Bankers Association, PO Box 676G, George Town, Grand Cayman.
Cayman Islands Law Society, PO Box 472, Grand Cayman.

COOK ISLANDS
Time Zone: GMT -10.5hrs (-11.5hrs April to October)

General

Political stability: Very good

Bank secrecy law: Yes

English Common Law: Yes

Company type: IBC

Exchange controls: Nil

Offshore revenue tax: Nil

Tax treaties: Nil

Owner disclosure: No

Language: English and Maori

Domicile migration: Yes

Offshore Banking: Yes

'Shelf' companies: Yes

Corporate Requirements

Minimum shareholders: 0

Minimum directors: 1

Corporate directors: Yes

Secretary required: Yes

Bearer shares: Yes

Local Requirements

Registered office/agent: Yes

Local directors: No

Directors disclosed: Yes
(but not made public)

Nominees permitted: Yes

Local Meetings: No

Trusts

Availability: Yes

Asset Protection Trusts: Yes

Perpetuity: Unlimited

COOK ISLANDS GENERAL INFORMATION

Taxes

There are no taxes and the Cook Islands do not have a double taxation treaty with any other nation.

International Company (IC) Features

A Company can be incorporated as an international company under the Cook Islands International Companies Act if its shareholders are non-residents of the Islands.

There are no minimum capital requirements and shares may be of no par value. Bearer shares may be issued and registered shares that have been duly paid may be exchanged for bearer shares. Shares may be designated in all major currencies.

There are statutory provisions that allow bearer debentures, conversion of bearer debentures to registered debentures, perpetual debentures and for the re-issue of redeemed debentures.

Only one director of an international company need be appointed and there is no obligation to appoint a resident director. It is obligatory to have a resident secretary who must be an officer of a registered trustee company. Additional secretaries may be appointed who need not be residents.

There is no requirement for members to hold meetings in the Cook Islands. All companies must lodge annual returns. Audited accounts are not required where there has been no invitation to the public and the members of the Company resolve at each annual general meeting that auditors should not be appointed. There is also provision for the Registrar to allow inspection by any person showing 'good cause' but these words have been narrowly construed in the context of the strict secrecy provisions of the legislation.

The Cook Islands legislation permits such flexibility, which enable a company to transfer its domicile:

a)   Companies incorporated in other jurisdictions to transfer their registration to the Cook Islands as international companies; and

b)   International companies to transfer their registration to other jurisdictions.

Principals or promoters of international companies may remain anonymous, as there is no obligation to disclose any details of beneficial ownership of shares. Such anonymity is further guaranteed by penal sanctions on any person who discloses information derived from an inspection of the records of the international company. The documents lodged with the Registrar of International Companies are only available for inspection by directors, members and debenture holders. All offshore entities are formed through a trustee company licensed in the Cook Islands to carry on the business of a trustee company.

Trusts

The Cook Islands perhaps has the most aggressive legislation regarding Asset Protection Trusts (APTs) with its attractive International Trusts Act.

Investor Protection

The Cook Islands does not have an investor's compensation scheme. Where there is a complaint against a trustee company in respect of which the complainant has a cause of action, then the court is the appropriate avenue of redress. Trustee companies are licensed and any complaint would be a factor the Monetary Board would consider in granting or renewing a licence. In respect to domestic solicitors, complaints should be lodged with the Chief Justice under the Law Practitioners Act and legal redress may be pursued through normal civil actions as well.

Location

The Cook Islands comprise 15 islands in the South Pacific Ocean between Tahiti in the East and Samoa and Tonga in the West. The main island is Rarotonga and the township of Avarua on that island is the administrative and commercial centre of the Cook Islands.

Government contacts: The Secretary, Cook Islands Monetary Board, PO Box 594, Avarua, Cook Islands. Tel: 00682 20 798. Fax: 00682 21 798

Other contacts: Tony Manarangi, Solicitor, PO Box 514, Avarua, Rarotonga, Cook Islands. Tel: 00689 23840. Fax: 00689 23843. Clarkes PC, Solicitors, PO Box 144, Avarua, Rarotonga, Cook Islands. Tel: 00689 24567. Fax: 00689 21567

CYPRUS
Time Zone: GMT +2hrs (+1hr April to October)

General

Political stability: Good	Bank secrecy law: Yes
English Common Law: Yes	Company type: Exempt
Exchange controls: Nil	Offshore revenue tax: 4.25%
Tax treaties: Many	Owner disclosure: Yes
Captive Insurance: Yes	Domicile migration: Doubtful
Offshore Banking: Yes	'Shelf' companies: Yes

Corporate Requirements

Minimum shareholders: 2

Minimum directors: 1

Corporate directors: No

Secretary required: Yes

Bearer shares: No

Local Requirements

Registered office/agent: Yes

Local directors: No

Directors disclosed: Yes

Local Meetings: No

Trusts

Availability: Yes

Asset Protection Trusts: Yes

Perpetuity: 100 years

CYPRUS GENERAL INFORMATION

Taxes

Offshore companies pay tax in Cyprus; shareholders are not subject to any Cyprus tax on dividends from such companies.

Cyprus has an excellent network of double taxation agreements, most of which are comprehensive and follow the OECD model. See below for a list of countries with which Cyprus has double taxation agreements.

Confidentiality

Cyprus has very strict regulations regarding secrecy and the disclosure of information. The Central Bank of Cyprus requires the identity of the beneficial owners of the shares in an offshore company to be disclosed to them.

However, each director, officer, or employee of the bank is bound to secrecy and is criminally liable for disclosing information to an unauthorised person. The bank employees are also required upon appointment to give an oath of secrecy that complete secrecy is maintained and information not be disclosed to any third party

Offshore Company Features

Companies registered in Cyprus, whose shares are directly or indirectly owned by foreigners and whose income is derived from sources outside Cyprus, qualify as 'offshore companies'.

An offshore entity registered and operating from Cyprus can take any one of three legal forms. It can be a limited liability company; a branch in Cyprus of a company registered abroad, or a partnership.

The Companies Law (as amended) which is almost identical to the United Kingdom's former Companies Act (1948), provides for private or public companies.

A private company, the preferred form for offshore planning, means a company which by its articles:

- Prohibits any invitation to the public to subscribe for its shares or debentures

- Limits the number of its members to 50 with a minimum of 2; and

- Restricts the right to transfer its shares and prohibits the issue of bearer shares.

Private companies can be further divided into exempt and non-exempt. The conditions, which must be fulfilled by a private company in order to be considered exempt, are that:

- No corporate body holds any of its shares or debentures, unless it is itself an exempt private company.

- no person other than the holder has any interest in the shares or debentures;

- The number of persons holding the company's debentures does not exceed 50;

- No corporate body is a director of the company.

An exempt private company enjoys the following advantages:

- It need not annex to the annual return submitted to the Registrar of Companies a copy of its financial statements;

- It may give loans and guarantees to its directors;

- It may appoint as its auditor a person who does not have the statutory qualifications to act as such.

Company shares must be denominated in Cyprus Pounds and cannot be issued in bearer form, only registered form. Should anonymity be required, nominees may hold the shares in trust for the beneficial owner.

It is possible to 'buy' ready-made 'Shelf' companies that have been inactive since incorporation. This may save time and expense in relatively simple instances but may prove to be more costly if substantial alterations need to be made to the memorandum and articles of association.

Meetings of the board of directors may be held in either Cyprus or abroad. If, from the tax planning point of view, it is important that the company be managed and controlled in Cyprus, then the majority of directors must be Cypriot residents.

Every company must have a registered office in Cyprus from the day it commences business or from the fourteenth day after its incorporation, whichever is earlier.

Cyprus Tax Treaties

Austria	Greece	Poland
Bulgaria	Hungary	Romania
Canada	India	Slovak Rep.
China	Ireland	Sweden
Czech Rep.	Italy	Syria
Denmark	Kuwait	Ukraine
Egypt	Malta	United Kingdom
France	Norway	United States
Germany	Yugoslavia	

Cyprus's role as a financial centre is greatly enhanced by the country's network of double tax treaties. Financial organisations which, as a result of their business activities, derive income in the nature of interest, dividends, capital gains, royalties, and other income of similar nature, may take advantage of Cyprus double tax treaties and local legislation incentives.

A Cyprus offshore company is subject to tax at just 4.25 percent. Thus interest, royalties, dividends and capital gains received in Cyprus are subject to reduced/exempt treaty rates. In fact, capital gains are not taxed in Cyprus and, depending on the country of source of the gain, exemption may also be achieved in that country as well. With the opening up of Eastern Europe, Cyprus is uniquely positioned because of its double tax treaties with the Eastern European countries. Cyprus' treaties with these countries, compared with those signed by Eastern Europe with other countries, are much more advantageous.

Double tax treaties and tax planning through treaty shopping is complex. Cyprus, with its network of double tax treaties with East and West, and its attractive offshore tax incentives, is high on the list of ideal tax planning locations.

Location

The island of Cyprus is located at the eastern end of the Mediterranean Sea south of Turkey, west of Syria, and north of Egypt.

Government contact: Central Bank of Cyprus International Division, P.O. Box 5529 Nicosia, Cyprus

DUBLIN
Time Zone: GMT

General

Political stability: Excellent

English Common Law: Yes

Exchange controls: Nil

Tax treaties: Yes

Language: English

Offshore Banking: No

Bank secrecy law: No

Company type: Exempt

Offshore revenue tax: Nil

Owner disclosure: Not publicly

Domicile migration: Doubtful

'Shelf' companies: Yes

Corporate Requirements

Minimum shareholders: 2

Minimum directors: 2

Corporate directors: No

Secretary required: Yes

Bearer shares: No

Local Requirements

Registered office/agent: Yes

Local directors: No

Directors disclosed: Yes

Local Meetings: No

Trusts

Availability: Yes

Asset Protection Trusts: No

DUBLIN GENERAL INFORMATION

Taxes

Non-resident companies are taxed solely on their income from Ireland, not on foreign income. See below for a list of countries with which Ireland has double taxation agreements.

Company Requirements

To obtain and maintain the tax-free status of the non-resident company, the ownership and centre of management must be based outside of Ireland and the company should not derive any income from within Ireland. All companies must file an annual return with the registry but not necessarily with the tax authority.

Although disclosure of the identity of the beneficial owner(s) is required, the address, domicile, nationality and other details are not needed. Moreover, the name(s) are not kept on public record and the beneficial owner(s) does not have to sign any documents but may appoint nominees.

Irish non-resident companies need only nominal capitalisation.

Location

Ireland is located in Europe to the west of Great Britain.

Ireland Tax Treaties

Australia	Israel	Portugal
Austria	Italy	South Africa
Belgium	Japan	South Korea
Canada	Luxembourg	Spain
Cyprus	Netherlands	Sweden
Denmark	New Zealand	Switzerland
Finland	Norway	United Kingdom
France	Pakistan	United States
Germany	Poland	Zambia

Investor Protection

The centre's regulatory body, the Central Bank of Ireland, is currently in the process of implementing the provisions of the European Union Investor Compensation Directive.

Government contacts: The Industrial Development Authority (IDA), IDA Ireland, Wilton Place, Dublin 2, Tel: (01) 6034000. Fax: (01) 6034040.

The Central Bank of Ireland, PO Box No 559, Dame Street, Dublin 2. Tel: (01) 6716666. Fax: (01) 6716561.

GIBRALTAR

Time Zone: GMT +1hr (GMT April to October)

General

Political stability: Very good

Bank secrecy law: No

English Common Law: Yes

Company type: Exempt

Exchange controls: Nil

Offshore revenue tax: Nil

Tax treaties: Nil

Owner disclosure: Not publicly

Language: English and Spanish

Domicile migration: Doubtful

Offshore Banking: Yes

'Shelf' companies: Yes

Corporate Requirements

Minimum shareholders: 2

Minimum directors: 1

Corporate directors: Yes

Secretary required: No

Bearer shares: Yes

Local Requirements

Registered office/agent: Yes

Local directors: No

Directors disclosed: Yes, but can be nominees

Local Meetings: No

Trusts

Availability: Yes

Asset Protection Trusts: Yes

Perpetuity: 100 years + 'wait-and-see' rule

GIBRALTAR GENERAL INFORMATION

* Gibraltar is part of the European Community by virtue of being a European territory of the United Kingdom.

Taxes

Non-resident owned and controlled companies incorporated in Gibraltar which do not trade, earn or remit income to Gibraltar are not liable to company taxation. There is no value-added tax and there are no exchange controls.

Gibraltar is not a party to any double taxation treaties and its tax authorities do not exchange or disclose information with or to any other tax authorities.

Gibraltar Companies

All Gibraltan companies must maintain a registered office in Gibraltar and the Registrar of companies must be notified of the address.

There must be at least one director who can be either an individual or a corporation and it is normal to appoint a company secretary who may also be an individual or a corporation.

There must be a minimum of two shareholders who may be individuals or corporations of any nationality and the maximum number of shareholders is 50.

The information relating to directors and shareholders must be filed at the registry and this information is available to the general public. Confidentiality can be achieved by the use of nominees.

A Gibraltar company may be of advantage to a parent company outside the European Community requiring an efficient investment vehicle into the European Community

Exempt Company

The main requirements for exempt status are that no Gibraltan or resident has a beneficial interest in the shares of the company and that the company does not trade or carry on business in Gibraltar.

The company can maintain office premises in Gibraltar for the purposes of transacting business with non-residents or with other exempt or qualifying companies.Exempt companies may issue bearer shares but are required to deposit the warrants in an approved bank. A secrecy provision in the Companies Ordinance prevents the disclosure of details concerning the beneficial owners of the company. It is also possible to achieve shareholder confidentiality by the use of nominee shareholders rather than bearer warrants.

Qualifying Company

A Gibraltar incorporated company or a registered branch of an overseas-incorporated company may apply for registration under the Income Tax (Qualifying Companies) Rules 1993.

Qualifying companies pay tax in Gibraltar at a rate negotiated with the Gibraltar authorities of between 0% and 18%.

The use of a qualifying company is often more advantageous than an exempt company due to anti-avoidance provisions in other jurisdictions.

Investor Protection

There is no investors' compensation scheme at present but Gibraltar will be implementing the Investors' Compensation Directive. Investors may complain directly to the Commissioner, Financial Services Commission. Gibraltar's regulatory system broadly matches that of the United Kingdom.

Location

Gibraltar is at the southern tip of the Iberian Peninsular at the junction of the Atlantic Ocean and the Mediterranean Sea.

Government contact: Mr Anthony Fisher, Finance Centre Development Director, Department of Trade and Industry, Europort, Gibraltar, Tel: (350) 74804. Fax: (350) 71406. Web Site: **www.gibraltar.gi/fsc**

GUERNSEY
Time Zone: GMT

General

Political stability: Excellent	Bank secrecy law: No
English Common Law: Yes (for Trusts)	Company type: Exempt
Exchange controls: Nil	Offshore revenue tax: Nil
Tax treaties: UK, Jersey	Owner disclosure: Yes
Language: English	Domicile migration: Doubtful
Offshore Banking: Yes	'Shelf' companies: No

Corporate Requirements

Minimum shareholders: 2

Minimum directors: 1

Corporate directors: Yes

Secretary required: Yes

Bearer shares: No

Local Requirements

Registered office/agent: Yes

Local directors: No

Directors disclosed: Yes

Local Meetings: No

Trusts

Availability: Yes

Asset Protection Trusts: No

Perpetuity: 100 years

GUERNSEY GENERAL INFORMATION

Taxes

No liability to income tax arises on a company that is granted exempt status for a particular year from the Guernsey tax authorities, provided that the only income derived by the company from Guernsey sources is interest on bank deposits. Otherwise, income is taxable at a rate of 20%. There are no value-added taxes (VAT) in Guernsey. Guernsey has double-taxation agreements with Jersey and the United Kingdom.

Exempt Company

Consent to form a Guernsey exempt company must be received from the Financial Services Commission and the following information must be supplied:

1) Names and addresses of the ultimate beneficial owners,

2) Character references relating to the ultimate beneficial owners,

3) The company's objectives,

4) The location where the central management and control is to be exercised,

5) The authorised and intended share capital, and

6) Whether the company will be used to avoid an existing United Kingdom tax

All Guernsey companies must maintain a registered office in Guernsey. There must be a minimum of one director who may be of any nationality and there is currently no prohibition on the appointment of corporations as directors. Usually, a company secretary is also appointed.

The share capital of the company may be denominated in any currency but must have a par value. Bearer shares are not permitted and there must be a minimum of two shareholders that may be of any nationality.

The details of the directors, company secretary and shareholders have to be filed and an annual return has to be filed confirming the same. There is no requirement to file audited accounts. This information is available to the public, however nominees may be engaged by corporations to shield the identity of the actual beneficial owners.

A Guernsey company must hold an annual meeting in Guernsey and shareholders may attend in person or vote by proxy. There is no requirement for the meetings of directors to be held in Guernsey.
Ready-made 'Shelf' companies are not permitted.

Bank Secrecy
English common law requires banks and their employees to maintain secrecy, but there is no specific local statutory law regarding confidentiality. Professionals cannot disclose information requested by a foreign court order.

Political and Economic Status
Guernsey is self-governing and not part of the United Kingdom but a British Crown dependency (although the UK is responsible for foreign relations and defence). Guernsey is not a member of the European Union, which means that it is not under obligation to harmonise tax or other legislation with other EU jurisdictions. Protocol No.3 of the UK's Treaty of Accession defines the island's relationship with the EU. Effectively, while benefiting from trade liberalisation within the EU, Guernsey has no EU obligations.

Investor Protection
The Collective Investment Schemes (Compensation of Investors) Rules 1998 provide for compensation for investors in Class A schemes of up to £5 million in any year. Subject to this limit, the maximum compensation payable

per investor is 90% of the first £50,000 and 30% of the balance up to £100,000 i.e. a total of £60,000. The Compensation Scheme is not a funded scheme; every licensee who is a designated manager, designated custodian/trustee or principal manager of an authorised Class A1 or A2 scheme is a participant under the Rules and liable for its share of the total payment due thereunder.

Should an investor have a complaint against a service provider in Guernsey that they are unable to resolve, they may contact the Commission. However, the Commission does not include amongst its functions that of an Ombudsman or an arbitrator and a complaint may ultimately be a matter for deliberation in the Courts. The Commission's role is to investigate a complaint to establish whether there are any matters of concern to the Commission as a supervisory and regulatory body. As such, when the Commission receives a complaint against an institution for which it has formal responsibility it may, if it considers appropriate, seek to establish whether the institution continues to satisfy the Commission's criteria of selectivity in remaining 'fit and proper'.

Location
Guernsey is one of the Channel Islands (including Alderney, Jersey, and Sark) located in the English Channel off the north west coast of France with the United Kingdom to the north.

Government contact: Guernsey Financial Services Commission, Valley House, Hirzel Street St. Peter Port, Guernsey, Channel Islands. Tel: 44 (0) 481-712706. Fax: 44 (0) 481-712010. Web Site: http://gfsc.guernsey.net

ISLE OF MAN
Time Zone: GMT

General

Political stability: Excellent	Bank secrecy: Strict Common Law
English Common Law: Yes	Company type: IBC, Exempt, Guaranteed/Hybrid, LLC
Exchange controls: Nil	Offshore revenue tax: Nil
Tax treaties: UK	Owner disclosure: No
Language: English	Domicile migration: Doubtful
Offshore Banking: Yes	'Shelf' companies: Yes

Corporate Requirements

Minimum shareholders: 1
Minimum directors: 2
Corporate directors: No
Secretary required: Yes
Bearer shares: Yes

Local Requirements

Registered office/agent: Yes
Local director: Depends
Directors disclosed: Yes
Local Meetings: No

Trusts

Availability: Yes
Asset Protection Trusts: No
Perpetuity: 80 years or Life + 21 years

ISLE OF MAN GENERAL INFORMATION

Taxes

Non-residents are only taxed on local source income. There are no death or estate duties, capital transfer or gift taxes, capital gains or wealth tax.

The Isle of Man is a low tax area with a standard rate of income tax for residents of 15% and a higher rate of 20%. Being present on the Island for six months in any tax year will make that person resident. Regular visits to the Island may also create a residence situation.

Other than a limited tax treaty with the United Kingdom, the Isle of Man does not have any double taxation treaties.

Isle of Man Companies

Isle of Man companies can take a variety of forms:

- Sole trader

- Limited partnership

- Unlimited partnership

- Branch of foreign corporation

- Private company

- Public company

The minimum number of directors for all companies is two and they must be individuals. There is no restriction regarding nationality of directors although at least one director of an Exempt or International Company must be a Manx resident. The names and addresses of the directors are lodged with the Registrar of Companies and are available to the general public.

The company secretary may be an individual or a corporation but the secretary of an Exempt or International Company must be suitably qualified (see below). Isle of Man companies are generally permitted to issue bearer shares but this is not permissible for Exempt and International companies.

Disclosure of ultimate beneficial ownership is not required and nominee holdings are permissible. Ready-made 'Shelf' companies are available for immediate use.

Exempt Companies

An Exempt company may be resident on the Island and is exempt from Manx taxation on offshore income and dividends upon the lodging of an application for exempt status and payment of an annual fee to the Registrar of Companies. The company must apply each year for the tax exemption.

In addition to the tax exemption on profits from abroad, an advantage of an Exempt company is that it may be resident on the island and, therefore, it is easier to demonstrate its tax residence.

These tax-exempt companies are managed from the island but must carry on business elsewhere and are limited to certain types of businesses.

An Exempt company must have at least two directors one of which must be a Manx resident, and the required resident company secretary must be suitably qualified as a solicitor or accountant, etc.

International Companies

The status of an International Company (IC) is that of a resident company for tax purposes which can be private or public and is subject to an income tax charge.

The Manx IC is quite similar to the Exempt Company except that legislation provides for a flexible rate of taxation that is chargeable according to the specific requirements of the IC. This can range from a minimum rate of 1% to a sliding scale maximum of 35% of assessable profits. The basis of assessment for arriving at the income tax chargeable can be negotiated to suit the Company's requirements.

No person resident in the Isle of Man can have a beneficial interest in a private IC, but public companies are permitted to have a small local shareholding of up to 5% without jeopardising IC status.

Non-Resident Companies

Non-resident companies are tax exempt on profits from abroad. Each year non-resident companies must file a non-resident declaration with the Registrar of Companies and pay an annual non-resident duty. Central management and control of these companies is exercised from a foreign location.

Resident Companies

Resident companies are used primarily for local business in the Isle of Man and are subject to income taxation rates up to 20 percent.

Political and Economic Status

The Isle of Man is not a member of the European Community, but enjoys a special relationship with the EC as set out in Protocol 3 to the UK's Treaty of Accession, which allows free trade in industrial and agricultural products with the Community. The Island is not part of the 1992 Single Market program, and, as a fiscally independent country, will not be subject to any harmonisation of taxes within the EC.

Investor Protection

The Isle of Man does have an investor's compensation scheme. If an investor has a complaint then, depending on the type of complaint, he or she should send a letter to the appropriate regulatory body, the Department of Consumer Affairs and the Courts.

Location

The Isle of Man is located in the centre of the Irish Sea equidistant from England, Ireland and Scotland.

Government contacts: Commercial Development Division, Treasury, Dept. 500, Government Offices, Douglas, Isle of Man. Tel: 44 (0) 1624 685755. Fax: 44 (0) 1624 685747.

Financial Supervision Commission and Insurance Pensions Authority, PO Box 58, 1-4 Goldie Terrace, Douglas, Isle of Man. Tel: (0) 1624 624487. Fax: 44 (0) 1624 629342.

JERSEY

Time Zone: GMT

General

Political stability: Excellent	Bank secrecy law: No
English Common Law: Yes (for Trusts)	Company type: Exempt, IBC
Exchange controls: Nil	Offshore revenue tax: 0-2%
Tax treaties: UK, Guernsey	Owner disclosure: Not publicly
Language: English	Domicile migration: Yes
Offshore Banking: Yes	'Shelf' companies: No

Corporate Requirements

Minimum shareholders: 2	
Minimum directors: 1	
Corporate directors: No	
Secretary required: Yes	
Bearer shares: No	

Local Requirements

Registered office/agent: Yes
Local directors: No
Directors disclosed: No
Local Meetings: No

Trusts

Availability: Yes

Asset Protection Trusts: No

Perpetuity: 100 years

JERSEY GENERAL INFORMATION

Taxes

There are no capital gains taxes, estate or inheritance duties nor are there any value-added taxes (VAT) or sales taxes.

Any company incorporated in Jersey that is neither an exempt company nor an IBC is an Income Tax Company and pays tax in Jersey on its worldwide income at a rate of 20%.

Jersey has full taxation treaties with Guernsey and the United Kingdom.

Jersey Companies

The two primary types of Jersey companies used for offshore financial planning are the Exempt Company and the International Business Company (IBC).

All Jersey companies must maintain a registered office in Jersey, the address of which must be given to the Financial Services Department.

Information regarding the beneficial owner(s) must also be submitted but is not disclosed to any other Jersey department or outside the government.

There must be at least two directors for a public company (one for private company) who must be individuals but may be of any nationality. A Company secretary must also be appointed.

The share capital of a Jersey company may be of any amount in any currency but bearer shares are not permitted. The minimum number of shareholders is two of any nationality.

Ready-made 'shelf' companies are not available, but in 1994 fast track incorporations were introduced, allowing formation of a Jersey company in less than two hours.

Exempt Company

Under the Jersey Company Law, an exempt company pays income tax only on Jersey source income other than interest arising from Jersey bank and building society deposits; that is, no tax on income arising outside the island.

An exempt company is not considered a resident in Jersey and no resident of Jersey may have any ownership interest in the company. However, locally resident nominees or trustees acting in those capacities for the benefit of non-residents may hold shares of an exempt company.

International Business Company (IBC)

An IBC is taxed on profits from international operations depending on the level of profit. IBCs are not allowed to trade locally.

Despite being owned by a non-resident, an IBC, unlike an exempt company, is resident in Jersey for tax purposes (except in the case of local branches of overseas companies) but pays Jersey income tax at reduced rates.

IBCs are of particular use for companies with international operations by providing a tax efficient structure for offshore activities such as import/export, international finance and insurance business and for reinvoicing (a dubious method of avoiding tax or duties on imports/exports).

IBCs are often used to exploit double taxation agreements with other countries which require that traders pay some tax in their jurisdiction. For example, a Scandinavian company may set up a Jersey IBC from which to run its treasury operations, enjoying double tax relief while paying only 2% local tax.

Jersey Trusts

Jersey trusts are used in a variety of tax and estate planning procedures for private clients and are often formed with an underlying Jersey company.

Although there are various types of trusts, the two most popular in Jersey are the fixed interest and the discretionary trusts.

With discretionary trusts, trustees have the choice as to whom to pay income and capital out of a predetermined list of beneficiaries. The person putting assets into the trust, the settlor or grantor, normally also gives the trustees a letter of wishes regarding disposal of the trust's income and assets. The attraction of a discretionary trust is that it allows flexibility to meet all the changing family circumstances, and can provide fiscal benefits as well.

In a fixed trust, the settlor predetermines disposal of the assets, including the entitlement of each beneficiary. The settlor can be named in the trust deed or can remain anonymous.

Set up fees and annual management costs depend on the trust firm chosen, the complexity of the trust deed and the amount of work involved in administering the trust.

Jersey is party to The Hague Convention on the Law Applicable to Trusts and on Their Recognition. This confirms the general international acceptability of Jersey trusts and the Jersey trust jurisdiction as well as the suitability and comprehensiveness of its trust legislation.

Trusts for non-residents are exempt from income tax on overseas income and on bank deposit interest arising in Jersey.

Investor Protection
Jersey does not operate an investor's compensation scheme.

Location
Jersey is one of the Channel Islands (including Alderney, Guernsey, and Sark) located in the English Channel off the north west coast of France with the United Kingdom to the north.

Government contact: Financial Services Commission, Cyril Le Marquand House P.O. Box 267, The Parade, St. Helier, Jersey JE4 8TZ, Channel Islands, Tel: +44(0) 534 603600

Fax: +44(0) 534 89155 Web Site: **www.itl.net/business/ofw/Jersey**

LIECHTENSTEIN

Time Zone: GMT +1hr (GMT April-September)

General

Political stability: Excellent	Bank secrecy law: Yes
English Common Law: Yes (for Trusts)	Company type: Exempt (Anstalt)
Exchange controls: Nil	Offshore revenue tax: Nil
Tax treaties: Austria	Owner disclosure: No
Language: German	Domicile migration: Yes
Offshore Banking: No	'Shelf' companies: Yes

Corporate Requirements

Minimum shareholders: 1

Minimum directors: 1

Corporate directors: Yes

Secretary required: No

Bearer shares: Yes

Local Requirements

Registered office/agent: Yes

Local directors: Yes

Directors disclosed: Yes

Local Meetings: No

Trusts

Availability: Yes

Asset Protection Trusts: Yes

Perpetuity: Unlimited

LIECHTENSTEIN GENERAL INFORMATION

Taxes

All individuals, trusts, and business entities domiciled, or possessing property or business premises, in Liechtenstein are subject to personal income tax, or corporate (profit) tax.

There is no Liechtenstein income tax levied against any company that is domiciled here if the company does not receive local source income.

Special tax rates apply to domiciliary and holding companies, foundations and trust. Liechtenstein's sole double taxation treaty is with Austria.

Liechtenstein Business Entities

Liechtenstein's company law permits the establishment of any type of legal entity found anywhere in the world. The two most popular categories of companies are holding companies and domiciliary companies.

The number of registered companies in Liechtenstein outnumbers its population by almost 2 to 1

Holding Companies

A holding company is defined in Liechtenstein as a company organised for the administration of capital or assets such as shares or bonds of one or more other companies. It may also own local or foreign property, including copyrights, licenses, patents, trademarks, etc.

Domiciliary Companies

A domiciliary company has only a corporate seat but no business activities in Liechtenstein. The most common corporate forms are the stock corporations Aktiengesellschaft (AG) and the Anstalt, a corporate/trust hybrid.

Anstalt

Also known as 'Establishment' or 'Establissement', this legal entity combines the best of corporate and trust features. It is unique to Liechtenstein and is very popular with those who want (and are willing to pay for) the utmost in financial secrecy.

Stiftung

The Stiftung is the Liechtenstein version of the trust and permits flexibility in its structure and purpose. If structured to the particular needs of the individual client, the foundation should not cause any problems with any home country tax authorities.

Treuunternehmen

Yet another hybrid corporate/trust concoction from the creative lawyers in Liechtenstein.

Bank Secrecy

Liechtenstein's secrecy laws are among the strictest in the world and Liechtenstein is not a party directly or indirectly to any exchange-of-information agreements.

Investor Protection

There is no fixed investor's compensation scheme. The respective agreement is up to every client and his professional partner. Complaints can be filed with the Government, with the Banking Commission, with the regular courts and with the Liechtenstein Associations of Professional Trustees or, if the complaint is against a lawyer, the Liechtenstein Bar Association.

Location

Liechtenstein is a small municipality of approximately 160 square kilometres (60 square miles) located between Austria and Switzerland.

Government contacts: Liechtenstein Government, Regierungskanzlei, FL-9490 Vaduz, Liechtenstein. A list of licensed trustees and attorneys can be obtained from the Government.

Other contacts: The Liechtenstein Association of Professional Trustees, Liechtensteinische Treuhandervereinigun, Postfach 814. FL-9490. Vaduz, Liechtenstein.

LUXEMBOURG
Time Zone: GMT +1hr (GMT April to September)

General

Political stability: Excellent	Bank secrecy law: Yes
English Common Law: No (Civil Law)	Company type: Holding (S.A.)
Exchange controls: Nil	Offshore revenue tax: Nil
Tax treaties: Many	Owner disclosure: No
Language: French, German, Luxembourgish	Domicile migration: Probable
Offshore Banking: No	'Shelf' companies: Perhaps

Corporate Requirements

Minimum shareholders: 2

Minimum directors: 3

Corporate directors: Yes

Secretary required: No

Bearer shares: Yes

Local Requirements

Registered office/agent: Yes

Local directors: No

Directors disclosed: Yes

Local Meetings: Yes

Trusts
Availability: No

LUXEMBOURG GENERAL INFORMATION

Taxes

There is no withholding tax on bank or bond interest and certain holding companies are exempt from tax on profits. See below for a list of countries with which Luxembourg has double taxation agreements.

Luxembourg Holding Companies

Luxembourg offers several legal business entities but the holding company is the most popular for offshore purposes because of its tax-free benefits.

There are two types of holding companies in Luxembourg and, though the difference is subtle, the tax results are not.

Holding companies are those companies whose sole object is to participate in other domestic or foreign companies through majority shareholdings. They may finance the companies in their group by investment or loan funds and may collect income from them consisting of dividends, loan interest, licensing fees, or patent royalties. These companies are exempt from corporation tax on profits, dividend and royalty income, and from capital gains tax and withholding tax on dividends payments, they may issue bearer shares and bonds.

The tax benefits and strategies available to Luxembourg holding companies are numerous and can be quite intricate. A knowledgeable international tax advisor should be consulted to reap the full potential of these investment vehicles.

Holding companies that also engage in direct commercial or industrial activity cannot benefit from the above tax advantages.

Luxembourg holding companies are often expressly excluded from the benefits of the Grand Duchy tax treaty network. Thus, while there is no Luxembourg withholding tax on certain outflows, dividend and interest income received by such companies are subject to maximum rates of withholding tax in the foreign source country.

Financial Secrecy

Luxembourg has banking secrecy laws tougher than Switzerland and is perhaps more lax to fraudsters and the criminal element. As a member of the European Community (EC), Luxembourg has been under some pressure to relax its secrecy laws, although it is unlikely to do so. In 1989, when the EC proposed to introduce withholding tax on all interest earned by EC residents and to increase co-operation between EC tax authorities, Luxembourg strengthened its banking laws. Well aware of the potential economic downfall with less stringent secrecy laws, the Luxembourg Prime Minister stated at the time that banking secrecy 'was not negotiable'.

Luxembourg banking services are relatively inexpensive, the tax regime is lenient, there is no tax withheld at the source, banks ask few questions of new customers, and numbered bank accounts are widely available. They will not, however, take in large quantities of cash from unknown customers. Secrecy protection is strongest for holding companies that need divulge only minimal information.

The legal system of the Grand Duchy also appears to hinder rather than assist anyone who seeks to uncover information about the source or destination of assets. Obtaining a court order in Luxembourg to freeze stolen assets is very difficult, unless full details of the account in question, including the names of its signatories, are provided.

There is an obvious drawback to such a "hands off" policy: the occasional scandal. For example, the headquarters of the failed Bank of Credit and Commerce International (BCCI) was based here.

Luxembourg Tax Treaties

Austria	Indonesia	Romania
Bahrain	Ireland	Singapore
Belgium	Israel	Slovak Rep.
Brazil	Italy	Slovenia
Bulgaria	Ivory Coast	South Africa
Canada	JapanSouth	Korea
China	Kuwait	Spain
Czech Rep.	Macau	Sweden
Denmark	Malta	Switzerland
Finland	Mauritius	Togo
France	Morocco	Ukraine
Gambia	Netherlands	United Kingdom
Germany	New Zealand	United States
Greece	Norway	Vietnam
Hungary	Poland	

Location

This small European nation borders Belgium, France and Germany.

Government contact: Institut Monetaire Luxembourgeois (IML), 63 Avenue de la Liberte, 22983 Luxembourg. Tel: 352 40 2929250. Fax: 352 492180

Other contacts: Association Luxembourgeoise des Fonds a Investissment (ALFI), 5 rue Aldringen, L-1118 Luxembourg. Tel: 352 223026

NETHERLANDS ANTILLES
Time Zone: GMT -4 hrs (-5 hrs April to October)

General

Political stability: Excellent

English Common Law: No (Dutch Civil)

Exchange controls: Nil

Tax treaties: Few

Language: Dutch/Papiamento (Spanish and English widely spoken)

Offshore Banking: Yes

Bank secrecy: Yes

Company type: Exempt (N.V.)

Offshore revenue tax: 2.4-3%

Owner disclosure: No

Domicile migration: Yes

'Shelf' companies: Yes

Corporate Requirements

Minimum shareholders: 1

Minimum directors: 1

Corporate directors: Yes

Secretary required: No

Bearer shares: Yes

Local Requirements

Registered office/agent: Yes

Local director: Yes

Directors disclosed: Yes

Local Directors Meetings: No

Local Shareholders Meetings: Yes

Trusts

Availability: No

NETHERLANDS ANTILLES GENERAL INFORMATION

Taxes

Offshore income from dividends and interest is taxable. Income from capital gains is tax exempt. Capital losses are not tax deductible.

Exempt Company Requirements

Companies formed under the laws of the Netherlands Antilles for the purpose of deriving income from foreign investments, trading, or other activities outside the Netherlands Antilles, and whose shares solely non-residents hold, are defined as 'offshore' companies.

The most common way of establishing a business in the Netherlands Antilles is in the form of a limited liability company called a 'naamloze vennootschap' (NV). The NV is a body corporate with shareholders and governed by the corporate law and its articles of incorporation (Articles of Association). The corporate law of the Netherlands Antilles is based on the Dutch corporate law but is not identical to it. The NV is to be formed by at least two founders, being either individuals or legal entities. Non-residents can establish a NV by proxy. It is possible for two residents of the Netherlands Antilles (either individuals or legal entities) to establish the corporation and subsequently transfer the shares to one person (individual or legal entity) resident or non-resident of the Netherlands Antilles. The NV can thus be owned by only one shareholder.

Obviously, ready-made 'Shelf' companies are available for immediate use.

The company must have one or several 'Managing Directors' who can be either an individual or a legal entity. At least one managing director must be a resident of the Netherlands Antilles. The Articles may also provide for a 'Board of Directors'.

The Netherlands Antilles corporate law does permit companies to issue bearer shares.

Tax Treaties of the Netherlands Antilles

Argentina	Norway	United Kingdom
Cuba	Suriname	Venezuela
Netherlands		

Location

The Netherlands Antilles consists of two groups of islands in the Caribbean Sea: the Leeward Islands group including Bonaire and Curacao about 33 miles off the north coast of Venezuela; and 500 miles to the northeast, the Windward Islands including St. Eustatius, St. Maarten, and Saba, which are about 100 miles east of Puerto Rico.

Government contacts: Bank van de Nederlandse Antillen (Central Bank), Breederstraat 1, Willemstad, Curacao, NA. Tel: 599 9 327000. Fax: 599 9 327590

Other contacts: Offshore Association of the Netherlands Antilles (VOB), Paviljoen Zeelandia, Suite I/J, Kaya WFG, (Jombi) Mensing 24, Curacao NA. Tel: 599 9 618111. Fax: 599 9 617948

SWITZERLAND

Time Zone: GMT + 1 hour (GMT April-September)

General

Political stability: Excellent	Bank secrecy law: Yes
English Common Law: No (Civil Law)	Company type: Holding (AG)
Exchange controls: Nil	Offshore revenue tax: Low
Tax treaties: Many	Owner disclosure: No
Language: German, French, Italian	Domicile migration: Yes
Offshore Banking: No	'Shelf' companies: No

Corporate Requirements

Minimum shareholders: 1	Registered office/agent: Yes
Minimum directors: 1	Local directors: Yes
Corporate directors: No	Directors disclosed: Yes
Secretary required: No	Local Meetings: No
Bearer shares: Yes	

Local Requirements

Trusts

Availability: No

SWITZERLAND GENERAL INFORMATION

Taxes

Switzerland has a federal structure of taxation whereby taxes are imposed concurrently by three separate authorities: federal government, cantons, and municipalities.

At the federal level, corporate tax amounts to 8.5% of profits. The tax rates vary greatly among the cantons, which are generally progressive. The maximum combined cantonal and municipal tax rate usually does not exceed 30% but there are exceptions.

Dividends paid by Swiss companies are subject to a 35% withholding tax that is credited against the taxes of resident shareholders or is refunded. It may be reduced or exempted under tax treaties for non-residents. A 35% withholding tax is generally levied on interest payments on bonds, debentures and bank deposits but can be reduced by the tax treaty system. See below for a list of countries with which Switzerland has double-taxation treaties. There is no withholding tax on interest income from foreign bonds issued in Switzerland.

Bank Secrecy

Swiss banks are required by law to maintain confidentiality about client affairs, under penalty of law. All clients and bank accounts benefit from this protection. An indiscreet bank employee can be imprisoned for up to six months, or fined up to Sfr50,000, for revealing client information without proper authorisation. Contrary to popular lore, accounts cannot be opened anonymously, whether numbered or regular.

The name of a client who opens any numbered account is known to a limited amount of bank executives. That is, a numbered account does not provide complete anonymity; it only limits the number of bank personnel who know the true identity of the account holder. The numbered account system was designed to reduce the risk of a bank employee inadvertently divulging unauthorised information about a client.

Under the treaty with the United States, Switzerland will disclose financial information when the US provides enough evidence that a crime, usually drug trafficking or money laundering, has been committed. Under current Swiss criminal law, anyone who knew or should have known that assets had a criminal origin and did not disclose the owner's identity would be imprisoned. However, mere allegations will not be sufficient and it must be crime under Swiss law. Tax avoidance is not a crime in Switzerland, but tax fraud and insider trading are.

Swiss Banking

Given the competence and integrity of Swiss private banking, investors choose to deposit funds in Swiss banks for several reasons other than secrecy.

These include the broad range of financial services offered, freedom from government interference, political stability, well known professionalism of Swiss bankers, and the ability to convert deposits to any stable currency as a hedge against inflation and currency fluctuations.

Switzerland's banks remain the cornerstone of its financial services industry, relying on low-taxes and strict banking secrecy laws to attract an international clientele, despite the recent embarrassing revelations about connections between the banks and Nazi Germany.

Swiss Companies

Swiss companies can issue either registered or bearer shares with nominal value. Incorporation takes approximately two weeks and the incorporators may be individuals or corporations. Directors and incorporators may be the same persons of any nationality. Nominee incorporators are permissible.

Swiss holding companies enjoy tax relief at the federal, cantonal and municipal levels. Despite the low tax rates, other jurisdictions continue to be more popular for establishing a holding company.

Switzerland Tax Treaties

Algeria	Ghana	Mexico
Argentina	Greece	Morocco
Australia	Hungary	Netherlands
Austria	Iceland	New Zealand
Belgium	India	Nigeria
Brazil	Indonesia	Norway
Bulgaria	Iran	Pakistan
Canada	Ireland	Poland
China	Israel	Portugal
C.I.S.	Italy	Romania
Czech Republic	IvoryCoast	Russia
Denmark	Jamaica	Singapore
Ecuador	Japan	Slovak Republic
Egypt	Kenya	South Africa
Finland	Lebanon	South Korea
France	Luxembourg	Spain
Germany	Malaysia	Sri Lanka
Sweden	Turkey	Uruguay
Tanzania	United Arab Emirates	Yugoslavia
Trinidad & Tobago	United Kingdom*	Zaire
Tunisia	United States	

* Extended to Anguilla, Antigua, Barbados, Belize, British Virgin Is., Dominica, Falkland Is., Gambia, Grenada, Malawi, Montserrat, Nevis, St. Christopher, St. Lucia, St. Vincent, Zambia, Zimbabwe.

Location

Switzerland is in central Europe and borders Austria, France, Germany, Italy and Liechtenstein.

Government contact: Eidgenossische Banken-Kommission (Federal Banking Commission) Marktgasse 37, 3001 Bern. Tel: 41 31 322 69 11. Fax: 41 31 322 69 26

Other contacts: Schweizerische Alagefondsverband (Swiss Investment Funds Association) Aeschenplatz 7, 4002 Basel. Tel: 41 61 295 93 93. Fax: 41 61 272 53 82

TURKS & CAICOS
Time Zone: +5 hrs (+4hrs April-October)

General

Political stability: Excellent

Bank secrecy law: Yes

English Common Law: Yes

Company type: Exempt

Exchange controls: Nil

Offshore revenue tax: Nil

Tax treaties: Nil

Owner disclosure: No

Language: English

Domicile migration: Yes

Offshore Banking: Yes

'Shelf' companies: Yes

Corporate Requirements

Minimum shareholders: 1

Minimum directors: 1

Corporate directors: Yes

Secretary required: Yes

Bearer shares: Yes

Local Requirements

Registered office/agent: Yes

Local directors: No

Directors disclosed: No

Local Meetings: No

Trusts

Availability: Yes

Asset Protection Trusts: Yes

Perpetuity: Unlimited

TURKS & CAICOS GENERAL INFORMATION

Taxes

The Turks & Caicos Islands (TCI) is a zero tax domicile with strict confidentiality laws. The TCI is not a party to any double-taxation agreements.

US Treaty

Limited Exchange of Information Agreement signed with the US Department of Justice under Mutual Legal Assistance Agreement which calls for exchange of information on tax fraud and laundering of funds but not offences which are purely tax-related.

Turks & Caicos Exempt Company Features

In addition to paying no tax on income, the exempt company's beneficial owners, directors, members, and officers do not have to be identified.

The company may purchase or own its own shares, at any par value in any currency. Only one shareholder is required for registered shares and bearer shares may be issued.

A director and a secretary are necessary but one person may hold both positions and be the sole shareholder. Nominee directors, officers, and shareholders are permitted. A registered office in TCI is required.

Incorporation takes only 2-3 days, but ready-made 'Shelf' companies are available for immediate use.

Investor Protection

Turks & Caicos does not have an investor's compensation scheme. Complaints may be forwarded to the Superintendent, Financial Services Commission.

Location

The Turks and Caicos Islands are located at the southern end of the Bahamas islands chain, east of Cuba and north of Haiti and the Dominican Republic.

Government contacts: Superintendent, Financial Services Commission, PO Box 173, Grand Turk, Turks & Caicos, British West Indies. Tel: 649 946 2937. Fax: 649 946 2557.

FURTHER CENTRE CONTACTS

Andorra

Time zone: GMT +1hr (GMT April to September)
Location: The tiny Principality of Andorra (464 sq km) lies between France and the Spanish region of Catalonia, towards the eastern end of the Pyrenees.
Language: Catalan
Contact: Ministeri de Finances, Govern d'Andorra, Carer Prat de la Creu 72-74, Andorra la Vella, Principat d'Andorra. Tel: 376 829345. Fax: 376 860962

Malta

Time Zone: GMT +1hr (GMT April to September)
Location: The Republic of Malta is situated in the Mediterranean 60m miles south of Italy.
Language: Maltese and English
Contact: Malta Financial Services Centre, Attard, Malta. Tel: 356 441155. Fax: 356 441188

Hong Kong

Time Zone: GMT +8hrs (+7hrs April to October)
Location: Situated on the south-east coast of China, Hong Kong principally comprises Hong Kong Island, Kowloon Peninsula and New Territories. Central District is the chief financial and business centre.
Language: Cantonese, Mandarin, English

Contacts: Securities and Futures Commission (SFC), 12th Floor, Edinburgh Tower, 15 Queen's Road, Central, The Landmark, Hong Kong. Tel 852 2840 9201. Fax: 852 2845 9553

Hong Kong Investment Funds Association (KHIFA), Room 808, Tak Shing House, 20 Des Voeux Road, Central, Hong Kong. Tel: 852 2537 9912. Fax: 852 2877 8827

Mauritius

Time Zone: GMT +4hrs (+3hrs April to October)

Location: Mauritius is an island in the Indian Ocean, 500 miles east of Madagascar.

Language: English, French, Creole

Contact: Mauritius Offshore Business Activities Authority (MOBAA), Deramann Tower, 30 Sir William Newton Street, Port Louis, Mauritius. Tel: 230 211 0494. Fax: 230 212 9459

Singapore

Time Zone: GMT +8hrs (+7hrs April to October)

Location: The island of Singapore is situated at the southern tip of the Malaysian peninsula and is connected to Malaysia by a causeway.

Language: Malay, English, Mandarin, Tamil

Contact: Monetary Authority of Singapore, MAS Building, 10 Shenton Way, Singapore 079117. Tel: 65 225 5577. Fax: 65 229 9697

Labuan

Time Zone: GMT +8hrs (+7hrs April to October)

Location: Labuan is a Malaysian Island in the South China Sea off the northwest coast of Borneo. Roughly equidistant from the financial centres of Bangkok, Kuala Lumpur, Singapore, Jakarta, Manila, Taipei and Hong Kong.

Language: Bahasa Malaysia and English

Contact: Labuan Offshore Financial Services Authority (LOFSA), Level 17, Main Office Tower, Financial Park, Jalan Merdeka, Labuan 87000, Malaysia. Tel: 60 87 408188. Fax: 60 87 413328

Vanuatu

Time Zone: GMT +11hrs (+10hrs April to October)

Location: Vanuatu comprises a chain of about 80 islands in the South Pacific, 500 miles west of Fiji.

Language: English, French, Bislama

Contact: Vanuatu Financial Services Commission, PO Box 23, Port Vila, Vanuatu. Tel: 678 22247. Fax: 678 22242

Source: Lipper Analytical

GLOSSARY

ADVERSE PARTY - a person whose beneficial interest in a trust would be adversely affected by the (non) exercise of power that he/she has with the trust.

AKTIENGESELLSCHAFT (AG) - a public company limited by shares.

ALIEN - sometimes referred to as a person who is not a citizen of the country in which he/she lives. See also non-resident alien and resident alien.

ALIEN CORPORATION - company incorporated under the laws of a foreign country regardless of where it operates. 'Alien Corporation' is often synonymous with the term 'Foreign Corporation'. However, 'foreign corporation' is also used in US State law to mean a corporation formed in a different state other from which it does business.

ANSTALT - a hybrid corporate/trust business entity of Liechtenstein.

ARM'S LENGTH TRANSACTION - a transaction entered into by unrelated parties, each acting in their own best interest and avoiding any semblance of conflict of interest. It is assumed that the prices in this transaction are the fair market values of the property or services being transferred. Transactions between certain family members and between a corporation and its subsidiaries are usually not considered at arm's length.

ARTICLES OF INCORPORATION - formal document prepared by individuals wishing to establish a corporation. They must file these documents with the authorities in the jurisdiction in which the corporation wishes to reside. Usually, the rules governing the company's internal management are stated in its bylaws.

ARTICLES OF PARTNERSHIP - formal document drawn up by partners stating significant and important aspects of the partnership. Items included are capital contributions, profit and loss ratios, name of the enterprise, duration of relationship, and partners' duties.

ASSET PROTECTION TRUST - often used to protect owner's assets from future creditors or liabilities.

BADGES OF FRAUD - conduct that raises a strong presumption that it was undertaken with the intent to delay, hinder or defraud a creditor.

BANK SECRECY CODE - legislation protecting investors against unauthorised disclosure of financial information to outsiders (including tax agents, lawyers, foreign and domestic courts, etc.). In some jurisdictions there are laws that protect confidentiality beyond Common Law and that expected of professional conduct. Criminal and civil penalties, that include fines and imprisonment, can result if bank officials and others with fiduciary responsibilities break this law.

BENEFICIARY - regarding Trusts, it is the party entitled to the benefits of the trust property.

BESLOTEN VENNOOTSCHAP (BV) - a private company with limited liability.

BUSINESS TRUST - is a voluntary association of investors who transfer assets to Trustees with the legal authority to manage the investments or business. The investors (holders of trust certificates) have limited liability.

CHARITABLE TRUST - a trust created to make one or more gifts to a charitable organisation.

CLOSED-END FUND - is a type of fund that has a fixed number of shares and is usually listed on a major stock exchange. They tend to have specialised portfolios of securities (stocks, bonds, etc.) with specific investment goals of income, capital gains, or a combination of these. Shares often sell at a discount from net asset value because the managers of closed-end funds are perceived to be less responsive to profit opportunities than open-end fund managers, who must attract and retain shareholders. Also unlike open-end mutual funds, closed-end funds do not issue and redeem shares on a continuous basis.

COMMON LAW - is a system of jurisprudence that originated in England and then passed on to its territories and former territories (e.g., Bahamas, Bermuda, Canada, Cayman Islands, Cook Islands, Gibraltar, Singapore, United States, et al.). It is based on judicial precedent (court decisions) rather than on legislative enactment (statutes).

COMPANY MIGRATION - (Company redomiciliation) some jurisdictions permit a foreign company to continue as a local company regardless of any contrary provisions in the laws of its previous domicile; may also permit a local company to continue in a foreign jurisdiction without dissolution.

CONTINGENT BENEFICIARY - receives benefits of a trust or estate only upon the occurrence of a certain event, such as the death of a named beneficiary.

CONTROLLED FOREIGN CORPORATION (CFC) - is described in the US Tax Reform Act of 1986 as a foreign corporation in which more than 50% of the combined voting power of all classes of voting stock, or the total value of its stock, is owned by US shareholders on any day during its tax year. A US shareholder is defined as a US citizen who owns 10% or more of the voting stock. US shareholders are required to include in their income their pro rata share of the Subpart F income of the CFC.

CORPORATE VEIL - the use of a corporation to disguise or protect a person's business and/or investment activities from scrutiny or assets from creditors.

CORPUS - (Latin for body) the assets or property in a trust, such as real estate, securities and other personal property, cash in bank accounts, and any other tangible or intangible property included by the grantor.

CREATOR - a person who creates a trust. Also see settlor and grantor.

CURRENT ACCOUNT - an offshore, personal savings or checking account.

CUSTODIAN - a bank, financial institution or other entity that has the responsibility to manage or administer the custody or other safekeeping of assets for other persons or institutions.

DEATH TAX - property tax imposed upon the death of the owner, such as an Inheritance or Estate tax.

DECLARATION OF TRUST - a document creating a trust; a trust deed.

DISCRETIONARY TRUST - a grantor trust in which the trustee has complete discretion as to whom among the class of beneficiaries receives income and/or principal distributions. There are no limits upon the trustee or it would cease to be a discretionary trust. The letter of wishes could provide some "guidance" to the trustee without having any legal and binding effects. Provides flexibility to the trustee and the utmost privacy.

DOUBLE TAXATION - internationally, the taxation of foreign income at the foreign source and taxed again by home country, domestically, taxation of earnings at the corporate level, then again as stockholder dividends.

DOUBLE TAXATION AGREEMENT - an agreement or tax treaty between two nations to reduce or eliminate the imposition of double taxation on the same income.

ESTATE TAX - tax paid to the government on a deceased person's assets that have been left to heirs. The estate pays the tax, not the recipients. Usually, no estate tax exists for property going from one spouse to another. In the US, the estate pays the tax, not the heirs.

EXPATRIATE - an individual who has lost citizenship of the home country but may still be subject to its tax laws. In the US, under certain circumstances such an individual is still subject to US taxation as a non-resident alien or to a special "alternative tax" rate, whichever is higher.

FIDUCIARY - person, company, or association holding assets in trust for a beneficiary. The fiduciary has the responsibility of investing the money wisely for the beneficiary's benefit. In some jurisdictions, it is illegal for fiduciaries to mismanage or invest the entrusted funds for their personal gain. Trustees, executors of wills and estates, and those who administer the assets of underage or incompetent beneficiaries are examples of fiduciaries.

FLIGHT CAPITAL - money that flows offshore and likely never returns.

FOREIGN TAX CREDIT - credit allowed against home income taxes for foreign taxes paid, or with tax sparing provisions in some tax treaties, against some foreign tax exclusions.

GENERAL PARTNERS - has unlimited liability. (1) All members of an ordinary partnership are General partners. (2) In a Limited partnership, at least one member is a General partner, and the other partners are Limited partners whose only liability is their capital contribution.

GIFT TAX - tax levied on the donor of property or money made without adequate legal consideration. It is based upon the fair market value of the property as of the date of transfer.

GMBH - a German form of a limited liability corporation.

HAGUE CONVENTION - of 1961, simplifies the process by which foreign documents from one country are authenticated for use in another country. Widely used in many European countries, it was not adopted by the US until 1981.

HIGH NET WORTH INDIVIDUAL (HNWI) - an individual with more than $1,000,000 in liquid assets to manage.

HOLDING COMPANY - a corporation organized for the purpose of owning stock in and controlling one or more other corporations.

IBC - a corporation. See International Business Company or exempt company.

INHERITANCE TAX - levied upon the cash or fair market value of property received through inheritance. This tax is suffered by the receiver of such assets and not by the estate, as in estate tax.

INTERFIPOL - International Fiscal Police. The tax crime counterpart to INTERPOL.

INTERNATIONAL BUSINESS COMPANY (IBC) - peculiar to offshore havens, and has the benefits of confidential ownership, low or no taxation, and ease of set up. IBCs must be owned or controlled by non-residents, and generally cannot carry on any local business or invest in local securities.

INTERNATIONAL FINANCIAL AND BANKING CENTRE (IFC) - a country identified as being a tax haven.

INTERNATIONAL TRUST - a Cook Islands term for a special type of an Asset Protection Trust (APT). Governed by the laws of the Cook Islands.

INTER VIVOS TRUST - (Living trust) a trust established between living persons.

INVESTMENT INCOME - income from dividends, interest, and royalties.

LETTER OF WISHES- guidance and a request to the trustee having no binding powers over the trustee. There may be multiple letters. They must be carefully drafted to avoid creating problems with the settlor or true settlor in the case of a grantor trust becoming a co-trustee. The trustee cannot be a "pawn" of the settlor or there is basis for the argument that there never was a complete denouncement of the assets. Sometimes referred to as a side letter.

LIMITED LIABILITY - the restriction of a person's potential losses to his/her contributions to the business, i.e., the absence of personal liability. Corporations, limited partnerships, and certain trusts enjoy this feature.

LIMITED LIABILITY PARTNERSHIP (LLP) - it is a partnership in which each partner is protected from the liabilities of the other partners.

LIVING TRUST - (Inter vivos trust) a trust established between living persons.

LOOPHOLE - generally a tax benefit which is over-generous to investors and deemed unnecessary by the tax authorities. Blatant loopholes are seen as legitimate targets for most tax authorities.

MONEY LAUNDERING - the illegal process of converting 'dirty' money (i.e., money obtained illegally, especially from drugs) into 'clean' (legal) assets.

MUTUAL LEGAL ASSISTANCE TREATY - an exchange of information treaty between the US and foreign jurisdictions. It generally covers the mutually recognised crimes but not all crimes.

NAAMLOZE VENNOOTSCHAP (NV) - a public company with limited liability.

NOMINEE - person or company, such as a bank official, bank or brokerage house, into whose name securities or other assets are transferred by agreement. In the offshore world they are used as company directors and shareholders, and to purchase assets to hide the identity of the true owner(s).

NOMINEE DIRECTORS - are frequently used to represent the company where the real beneficial owner does not live in the tax haven or does not want his name appearing on company documents.

NOMINEE SHAREHOLDER - sometimes used to hold shares on behalf of, or for, the real beneficial owner who wishes to remain anonymous.

NRA - non-resident (of the US) alien. Not a US person as defined under the Internal Revenue Code (IRC).

OFFSHORE FUNDS - investment funds established offshore to avoid restrictive regulations and/or high costs.

OPEN-END FUND - one that issues new shares and buys shares back on demand. There is no fixed number of shares outstanding, which distinguishes it from Closed-end mutual funds.

PROBATE - the necessary procedures to prove the legal validity of a will.

RESIDENT ALIEN - an individual who has permanent resident status but has not been granted citizenship.

REVERSIONARY INTEREST - an interest in property that although transferred to another by a decedent, may return to the decedent's estate or is further transferred to a third party according to the wishes of the deceased.

REVOCABLE TRUST - a trust whereby the Grantor still exercises control over the trust's assets.

SETTLE - to create or establish an offshore trust.

SETTLOR - individual who establishes a trust for Beneficiaries for the purpose of conserving or protecting assets for the settlor or other beneficiaries. Also known as the Donor or Grantor.

SHELF COMPANY - Company that has been legally established with the proper authorities before purchase by an investor. Between establishment and purchase, the company has been sitting "on the shelf". The shares are usually in bearer form and the names of the directors and officers are substituted when the company is taken "off the shelf" and becomes active.

SHELL CORPORATION - term applied to companies set up as fronts to conceal fraudulent activities such as tax evasion. Usually has no significant assets or operations.

SITUS OR SITE - the situs is the domicile or dominating or controlling jurisdiction of the trust. It may be changed to another jurisdiction, to be sited in another country or US State.

SOCIETE ANONYME (SA) - a limited liability corporation established under French law. Requires a minimum of seven shareholders. In Spanish speaking countries, it is known as the Sociedad Anonima. Important characteristic of both is that the liability of the shareholder is limited up to the amount of their capital contribution.

SPARBUCH - an Austrian numbered savings account.

STAMP TAX - levied on the issuance or transfer of legal documents and or financial securities in some jurisdictions.

STATUTE OF LIMITATIONS - the deadline after which a party claiming to be injured by the settlor may (should) no longer file an action to recover his or her damages.

STATUTORY - that which is fixed by statutes, as opposed to Common Law.

STIFTUNG - a Liechtenstein form of private foundation.

TAX AVOIDANCE - payment of the least tax possible by using legal tax planning opportunities. Tax avoidance should not be construed as betraying or disregarding a public or patriotic duty. In fact, a majority of court decisions have supported this assertion.

TAX EVASION - not paying taxes legally due a governmental agency. Some examples include: deliberately understating the income; failure to file the proper documents; failure to pay tax when due; intentionally making a false oral or written statement to the tax authorities; or wilfully conspiring, by aiding or advising, in the preparation of false oral or written tax statements.

TAX HAVEN - a jurisdiction providing significant tax breaks to individuals or corporations. As described by the OECD: "a jurisdiction actively making itself available for the avoidance of tax which would otherwise be paid in relatively high tax countries". They also may have flexible company and trust legislation.

TAX HOLIDAY - tax benefits for a specified period of time to encourage and stimulate business investment in a particular area.

TAX SHELTER - investments that can protect or defer ("shelter") a portion of income from taxes.

TAX TREATY (AGREEMENT) - bilateral income tax treaties are generally to reduce or eliminate the possibility of "double taxation" of income or on dividends. May also contain provisions for an exchange of tax information.

TESTAMENTARY TRUST - created by a will and not, as the Inter Vivos or Living trust, during the lifetime of the Settlor.

TRANSFER TAX - levied upon the transfer or sale of a security or property; based on selling price or par value.

TRUST - agreement in which the trustee takes title to property (called the corpus) willingly donated by the grantor (donor) to protect, manage, or invest it for either the grantor or the trust's beneficiary. Trusts may be revocable or irrevocable. Trusts may form and own companies that in turn may be used to hold the trust property. (Because there is no requirement to file the trust instrument, the true beneficial ownership of the company can remain totally confidential.)

TRUSTEE - holds and administers property of a trust for the benefit of another (beneficiary).

TRUST PROTECTOR - a person appointed by the settlor to oversee the trust on behalf of the beneficiaries. In many jurisdictions, local trust laws define the concept of the trust protector. Has veto power over the trustee with respect to discretionary matters but no say with respect to issues unequivocally covered in the trust deed. Trust decisions are the trustee's alone. Has the power to remove the trustee and appoint trustees. Consults with the settlor, but the final decisions must be the protector's.

UNLIMITED LIABILITY - liability of owner(s) not limited to the owner's investment, but perhaps to personal property. Stockholders usually have limited liability; that is, they risk their investment in the enterprise but not their personal assets.

VOTING TRUST - a trust where two or more shareholders transfer their shares to two or more trustees who then have the voting rights. The depositing shareholders still receive any dividends.

WITHHOLDING TAX - (1) Deductions by a broker or agency from dividends to foreign investors. (2) Deductions by an employer from employee salaries for the payment of income taxes.

Source: Bloomberg/International Investment

USEFUL ADDRESSES

Financial Advice:

Institute of Financial Planning, Whitefriars Centre, Lewins Mead, Bristol, BS1 2NT, UK. Tel: 4 (0) 117 930 4434. Fax: 44 (0) 117 929 2214

Performance Statistics:

Lipper Analytical Services, Vector Court, 143-145 Farringdon Road, London UK EC1R 3AD, UK Tel: 44 (0) 171 520 2100. Fax: 44 (0) 171 833 4861

Standard & Poor's Micropal, Commonwealth House, 2 Chalkhill Road, London W6 8DW. UK. Tel: 44 (0) 181 938 7100. Fax: 44 (0) 181 741 0929

HSW, London & Manchester House, Park Green, Macclesfield, Cheshire, SK11 7NG, UK. Tel: 44 (0) 1625 511311. Fax: 44 (0) 1625 616616

Contributors:

City Financial Publications, publishers of *International Investment* and the Offshore Fund Total Expense Rahon (TER) Survey 1998, 7 Air Street, London W1R 5RJ. Tel: 44 (0) 171 439 3050

Old Mutual International, Victory House, Prospect Hill, Douglas, Isle of Man IM1 1QP. Tel: 44 (0) 1624 677877. Fax: 44 (0)1624 616616

Albany International, St Mary's, The Parade, Castletown, Isle of Man, IM9 1RJ, UK. Tel: 44 (0)1624 823262. Fax: 44(0)1624 822560

Arthur Andersen, 1 Surrey Street, London, WC2R 2PS (UK). Tel: 44 (0) 171 438 3000.

Portfolio Fund Managers, 64 London Wall, London EC2M 5TP (UK). Tel: 44 (0) 171 6380808

Forsyth Partners, 18 Barclay Road, Croydon CR0 1JN (UK). Tel: 44 (0)181 649 94440. Fax: 44 (0) 181 649 9441

Ogier & Le Masurier, PO Box 404 Whiteley Chambers, Don Street, St Helier, Jersey JE2 9WG. Tel: 44 (0)1534 504000. Fax: 44(0)1534 35328

Guinness Flight, 5 Gainsford Street, London, SE1 (UK). Tel: 44 (0) 171 522 2100

Argyll Investment Management, 60 Queen Anne Street, London W1 (UK). Tel: 44 (0) 171 935 8125. Fax: 44 (0) 171 935 9194

Bloomberg, City Gate House, 39 Finsbury Square, London EC2 (UK). Tel: 44 (0) 171 330 7500. Fax: 44 (0) 171 392 6000

Moneyfacts, North Walsham, Norfolk, NR28 OBD (UK). Tel: 44 (0) 1692 500765. Fax: 44 (0) 1692 500865

Centre for the Study of Financial Innovation (CSFI), 18 Curzon Street, London W1 (UK). Tel: 44 (0) 171 493 0173.

Regulatory:
Financial Services Authority, Gavrelle House, 2-14 Bunhill Row, London, EC1Y 8RA. Tel: 44 (0)171 638 1240. Fax: 44 (0)171 382 5900.

Tax:
Inland Revenue, South West Wing, Bush House, Strand, London WC2, UK. Tel: 44(0) 171 438 6420.

INDEX

Abbey National   91

absolute interest   61

acceptability concept   47

accumulator fund   40, 112

addresses   221-222

adverse party   209

aktiengesellschaft (AG   209

Albany International   98, 101-102

Albany Life International   59

alien   209

alien corporation   209

Altajir Bank   85

Andorra   206

anstalt   193, 209

appointor   62

appraising funds,   139

arbitrage   123

arm's length transaction   209

Arthur Andersen   50

articles of incorporation   210

articles of partnership	210
asset allocation	140, 142
Asset Protection Trusts (APTs)	64, 65, 75, 168, 210
assets protection	66
Association of Investment Advisers and Financial Planners	51
Badges of fraud	210
Bahamas	13, 23, 55, 65, 76, 78, 151-153
Bahrain	51
Banco Totto & Acores	91
Bank of Credit and Commerce International (BCCI)	196
bank secrecy	17, 193, 196, 201-202, 210
banking	15, 23, 83-95
Banque Internationale à Luxembourg	32-33
Barbados	42, 76, 153-155
Barclays Bank	106
Baring Asset Management	32
Barings	106
Belize	76
benchmarks	126
beneficiaries	60-61, 210
Bermuda	65, 75, 155-157
besloten vennootschap (BV)	210
Bilonline	33

Blackstone Franks International	51
Bloomberg	125
Bloomberg/International Investment	149
British Virgin Islands	23, 74, 76, 79, 157-161
building societies	87
business trust	210
Canada	23, 55
capping	134, 138, 139
Caribbean area	23
case studies	19-22, 117-119
Cater Allen	89
Cayman Islands	13-14, 15, 23, 30, 55, 65, 75, 76, 83, 84, 85, 162-165
Centre for the Study of Financial Innovation (CSFI)	28
Certified Financial Planner (CFP)	51
Certified Financial Planner Board of Standards	52
Channel Islands	see Guernsey & Jersey
charges	133-138
charitable trust	211
Charles Schwab	29-30
Charlton, Regina versus	46-48
Citibank	90
Citibank Financial Communications	28
civil law system	55-59
Clerical Medical International	97, 102

closed-ended funds	108-109, 211
closed-ended investment companies	110
common law system	55-59, 211
company formation	71-81
company migration	211
confidentiality	13, 71, 73, 78, 80, 84
contingent beneficiary	211
contingent interest	61
controlled foreign corporation (CFC)	211
Cook Islands	65, 166-169
Corbin, Giles	45-48
corporate veil	212
corpus	212
costs of offshore companies	73-74, 80
creator	212
Credit Suisse	90
criminal transactions	16-17
crossborder migration	74
cumulative performance basis	125
current account	212
custodian	212
custodian handling charges	138
Cyprus	23, 42, 65, 169-173
Dead settlor provision	57
death tax	212
decision making	17-19
declaration of trust	212

deed of settlement······································60

deed of trust··60

deposit accounts··40, 87

derivatives···14

disclosure problems·····································136

discreet performance analysis···························125-126, 140

discretionary trust····································212

distance selling directives·····························28

distressed securities···································123

distributing fund status································116

distributor funds·······································40, 112

domicile status···41

double taxation agreements······························212, 213, see also

    tax treaties

Dublin···23, 32, 110, 116, 174-176

Eastern Europe···173

economic instability····································11

economic stability······································79

emerging markets··124

estate planning···66

estate tax···213

Euro currency··89

European Union Investor Compensation Directive 175

exchange control·······································13, 80

exempt companies·······································75

expatriates··16, 19-20, 39, 49-5, 87,

    213

expense limitation policy	138
expense ratio survey	137-138
Fairness concept	47
family assets	65
fee multiplier	136
fiduciary	213
financial advice	49-54
Financial Planner Standards Council	52
Financial Planning Association	52
Financial Services Authority (FSA)	18, 105
Fixed Capital Investment Company (FCIC)	110
flight capital	44, 213
flight clause	68
fonds communs de placement (FCP)	109
forced heirship	56-57
foreign earnings deduction (FED)	39
foreign tax credit	213
Fosyth, Paul	139-142
France	51, 56, 75
fraud	16
fund of funds	113-114
fund performance	127-133, 143-147
fund performance, charges	133-138
fund pricing	111-113
fund size	141-142

G T Global	28-29
gearing	120, 121-122
general partners	213
geography	23, 68
George Soros' Quantum Fund	124
Germany	56, 75
Gibraltar	65, 74, 87, 176-179
gift tax	214
Global Asset Management	30-31, 105
glossary	209-220
gmbh	214
Guernsey	15, 23, 55, 74, 83, 85, 87, 97, 116, 179-182
Guinness Flight	33-34
Guinness Mahon Bank	85
Hague Convention	69, 214
Hansard Group	29
hedge funds	113, 119-124
Hedge Funds (book)	119
high net worth individual (HNWI)	214
Hills, Richard	119
Hills, Victor	119-124
holding company	214
Hong Kong	13, 38, 51, 76, 206-207
HSW	125

IBC	214
information	25, 27-35
ING	106
inheritance tax	214
innovative investment	14
Institute of Financial Planning (IFP)	52
insurance bonds	40-41
intellectual rights	73
inter vivos trust	215
Interfipol	214
International Business Companies or Corporations (IBCs)	23, 74, 76-78, 152, 158-160, 189, 214
International Certified Financial Planning Council	51
International Financial and Banking Centre (IFC)	214
international trust	215
internet	27-35
investment income	215
investment management	97-103
investment team	142
investment trusts	110
investor protection	18
investors	17
Ireland	see Dublin
Isle of Man	29, 73, 87, 89, 97, 183-187

Jardine Fleming	106
Jersey	15, 23, 55, 73, 76, 83, 84, 85, 87, 91, 97, 110, 116, 187-191
Labuan	207
legal advice	45-48
legal systems	55-59
legal systems	55-59
legality	15-17
letter of wishes	215
leverage	129
Liechtenstein	191-194
life assurance companies	97
life assurance policies	49, 100
life interest	61
life tenant	61
limited duration company (LDC)	75
limited liability	215
limited liability companies (LLC)	75-76
limited liability partnerships (LLP)	120, 215
limited partnerships	110
Lipper Analytical	106, 125
Lipper Overseas Fund Table (LOFT)	127
Lipper, Michael	127-133
liquidity	141-142
living trust	215
Lloyds Bank	91

loan facilities	14
long/short equity	121-122, 123
loophole	215
Luxembourg	13, 23, 32-33, 83, 97, 109, 110, 116, 194-197
Luxembourg holding companies	195
Luxembourg unit trust	109
Macro funds	124
Madeira	23
Malta	206
managed portfolio bonds	98
management fees	137
market neutrality	122
Mauritius	23, 207
Meiklejohn, Steven	45-48
merger arbitrage	123
Merrill Lynch	90, 106
Mexico	75
Micropal	106, 125
mid cap	139
migration	78
mobile employees	39
money laundering	16, 17, 84, 161, 215
monitoring performance	125
mortgages	91-92
multi-currency accounts	88-99
multi-currency banking	86

multi-manager fund	114
Murray Noble	51
mutual funds	110
Mutual Legal Assistance Treaties	16, 215
Naamloze vennootschap (NV)	199, 216
nationality status	15
Netherlands Antilles	198-200
neutral investment home	13
new funds	140
Nickerson, Kira	28
nil income gain	117
nominee	216
nominee directors	72, 216
nominee shareholders	71, 216
non-resident alien (NRA)	216
Offshore banks	15, 23, 83-95
offshore banks, list	94-95
offshore centres	
definition	13-15
information	149-208
selecting	11-25
see also country names	
offshore companies	71-81
costs	73-74, 80
establishing	78-79
reasons for establishing	72-73

selecting jurisdiction	79-80
types and uses	74-78
offshore deposit accounts	40, 87
offshore funds	105-142
definition	216
selection	138-142
structures	107-111
types	40
offshore investment funds	23
offshore life assurance	49, 100
offshore products	97-103
offshore trading	72
offshore trusts	see trusts
Old Mutual	31-32
open-ended funds	107, 108, 110, 216
open-ended investment company (OEIC)	110
opportunistic fed income	123
Panama	76
performance analysis	125-126
performance fees	135
personal portfolio bonds (PPBs)	100
political instability	11
political stability	68, 79
pooled investment vehicles	120
portfolio bonds	98
Portugal	51
potential interest	61

pre-marital agreements	66
pricing methods, funds	112-113
private banking	89-90
probate	66, 216
property abroad	91
protector	63, 69
publicity, lack of	18
Reasons for offshore investing	11-12
Regina versus Charlton	46-48
regulatory regime	14
relationship manager	89-90
remaindermen	61, 64
remittance basis	39
residence status	15, 41
resident alien	216
retiring abroad	38
reversionary interest	61, 216
revocable trust	217
risks	14, 140
Robeco	106
roll-up funds	40, 112, 115-119
Romas, Christina	127-133
Royal and Sun Alliance	97
Royal Bank of Scotland International	83, 85, 91
Royal Skandia	97

Scottish Life International	98, 100-101
Scottish Provident International	97, 101
Scottish Widows International	100
Securities and Exchange Commission (SEC)	15, 18, 105
Securities and Investments Board (SIB)	18, 105
Sedgwick Financial Services	50
selection of offshore centre	22-25
Serious Fraud Office (SFO)	35
set up costs	135
settle	217
settlor	60, 217
share classes	111-112
share dealing	92-94
shelf company	217, 218
shell corporation	218
short selling	120-122, 124
short-squeezed	121
Singapore	13, 207
situs or site	217
small cap	139
sociedad anónima (SA)	217
société anonyme (SA)	217
Société d'Investissement à Capital (SICAV)	110
South Africa	38
Spain	21, 49, 51, 52
sparbuch	217
stamp tax	218
statistics providers	106, 125

statute of limitations	218
status	
domicile	41
residence	41
statutory	218
stiftung	193, 218
stock exchanges	109
Sun Life International	98
Sweden	19
Switzerland	13, 17, 42, 51, 84, 200-204
Tax	37-48
authorities	17
avoidance	17, 45-48, 218
declaration	15
efficiency	13, 28
evasion	16-17, 45-48, 218
haven	13, 218
holiday	218
liability	37
mitigation	46
offshore companies	72
planning	11-12, 22, 37, 66
regimes	13
remittance basis	39
shelter	219
United States	16, 20

tax status

    investors   38-40

    products   38-40

tax treaties   41-42, 172-173, 175, 197, 203, 219

tax-free growth   37

TER capping   138

testamentary trust   219

Timberlake, Richard   133-138

total expense ratio (TER)   133-138

total return   125

Towry Law International   51

Trading Places (film)   113

transfer tax   219

treaty shopping   42-44, 158

treuunternehmen   193

trust concept   55

trust deed   60

trust protector   219

trustees

    appointment   62

    death   63

    definition   219

    duties   61-62

    number   61

    retirement   63

trusts	55-69
absolute	63-64
asset protection	64, 65
changing trust deeds	67
charitable	64
definition	59-60, 219
discretionary	64
fixed interest	63-64
flexible	64-65
flight clause	68
Hague Convention	69
length	67
life interest	64
operations	60-63
power of appointment	64
protector	69
purpose trusts	64
reasons for establishing	65-67
selecting an offshore jurisdiction	67-69
spendthrift clause	69
types	63-65
valid investments	67
Turks and Caicos Islands	75, 76, 204-206
Umbrella funds	110, 114-115, 135
Undertaking for Collective Investment in	
Transferable Securities (UCITS)	110-111
unit-linked whole life assurance	98

United States of America

    citizens   16, 31

    common law   55

    offshore funds not sold   15, 31

    payments to informers   45

    taxation   16, 20

unlimited liability   219

Vanuatu   208

Variable Capital Investment Corporation (VCIC)   110

volatility   140-141

voting trust   220

W T Fry   50-51

websites   28-35, 125

Wells, David   19-22

withholding tax   220

Woolwich International   91

Yield   125